Springer Briefs in Population Studies

Population Studies of Japan

Editor-in-Chief
Toru Suzuki, Chiba, Japan

Series Editors
Shinji Anzo, Tokyo, Japan
Hisakazu Kato, Tokyo, Japan
Noriko Tsuya, Tokyo, Japan
Kohei Wada, Tokyo, Japan
Hisashi Inaba, Tokyo, Japan
Minato Nakazawa, Kobe, Japan
Jim Raymo, New Jersey, USA
Ryuichi Kaneko, Tokyo, Japan
Satomi Kurosu, Tokyo, Japan
Reiko Hayashi, Tokyo, Japan
Hiroshi Kojima, Tokyo, Japan
Takashi Inoue, Tokyo, Japan
Toshihiko Hara, Sapporo, Japan

The world population is expected to expand by 39.4% to 9.6 billion in 2060 (UN World Population Prospects, revised 2010). Meanwhile, Japan is expected to see its population contract by nearly one third to 86.7 million, and its proportion of the elderly (65 years of age and over) will account for no less than 39.9% (National Institute of Population and Social Security Research in Japan, Population Projections for Japan 2012). Japan has entered the post-demographic transitional phase and will be the fastest-shrinking country in the world, followed by former Eastern bloc nations, leading other Asian countries that are experiencing drastic changes.

A declining population that is rapidly aging impacts a country's economic growth, labor market, pensions, taxation, health care, and housing. The social structure and geographical distribution in the country will drastically change, and short-term as well as long-term solutions for economic and social consequences of this trend will be required.

This series aims to draw attention to Japan's entering the post-demographic transition phase and to present cutting-edge research in Japanese population studies. It will include compact monographs under the editorial supervision of the Population Association of Japan (PAJ).

The PAJ was established in 1948 and organizes researchers with a wide range of interests in population studies of Japan. The major fields are (1) population structure and aging; (2) migration, urbanization, and distribution; (3) fertility; (4) mortality and morbidity; (5) nuptiality, family, and households; (6) labor force and unemployment; (7) population projection and population policy (including family planning); and (8) historical demography. Since 1978, the PAJ has been publishing the academic journal *Jinkogaku Kenkyu* (The Journal of Population Studies), in which most of the articles are written in Japanese.

Thus, the scope of this series spans the entire field of population issues in Japan, impacts on socioeconomic change, and implications for policy measures. It includes population aging, fertility and family formation, household structures, population health, mortality, human geography and regional population, and comparative studies with other countries.

This series will be of great interest to a wide range of researchers in other countries confronting a post-demographic transition stage, demographers, population geographers, sociologists, economists, political scientists, health researchers, and practitioners across a broad spectrum of social sciences.

Reiko Hayashi

Demographic Change and Policy Responses

The Case of Japan

Reiko Hayashi
Director-General
National Institute of Population and Social Security Research
Tokyo, Japan

ISSN 2211-3215   ISSN 2211-3223   (electronic)
SpringerBriefs in Population Studies
ISSN 2198-2724   ISSN 2198-2732   (electronic)
Population Studies of Japan
ISBN 978-981-13-0459-0   ISBN 978-981-13-0460-6   (eBook)
https://doi.org/10.1007/978-981-13-0460-6

© The Author(s), under exclusive license to Springer Nature Singapore Pte Ltd. 2025

This work is subject to copyright. All rights are solely and exclusively licensed by the Publisher, whether the whole or part of the material is concerned, specifically the rights of translation, reprinting, reuse of illustrations, recitation, broadcasting, reproduction on microfilms or in any other physical way, and transmission or information storage and retrieval, electronic adaptation, computer software, or by similar or dissimilar methodology now known or hereafter developed.
The use of general descriptive names, registered names, trademarks, service marks, etc. in this publication does not imply, even in the absence of a specific statement, that such names are exempt from the relevant protective laws and regulations and therefore free for general use.
The publisher, the authors and the editors are safe to assume that the advice and information in this book are believed to be true and accurate at the date of publication. Neither the publisher nor the authors or the editors give a warranty, expressed or implied, with respect to the material contained herein or for any errors or omissions that may have been made. The publisher remains neutral with regard to jurisdictional claims in published maps and institutional affiliations.

This Springer imprint is published by the registered company Springer Nature Singapore Pte Ltd.
The registered company address is: 152 Beach Road, #21-01/04 Gateway East, Singapore 189721, Singapore

If disposing of this product, please recycle the paper.

Preface

Population policy refers to efforts to influence the quality and quantity of a population through its three key demographic components—births, deaths, and migration—with the aim of achieving an optimal population. While the term was once narrowly synonymous with pronatalist policy, its proper definition covers a vast array of social, economic, and public policies related to family, health, employment, urban development, and border control, among others. This booklet examines how the Japanese people perceived population through data and responded to its changes in modern Japan, from the Meiji period (beginning in 1868) through the Taisho, Showa, Heisei, and Reiwa eras. Over these 150 years, the Japanese people have sometimes responded correctly based on data and other times have failed to respond, being unaware of the population change, either due to a lack of data, insufficient analytical skills, or an unwillingness to confront reality or take action.

This booklet is composed of three chapters on fertility, mortality, and migration. Each chapter details how population data was collected and how contemporary stakeholders—such as researchers, academics, and policymakers—perceived and responded to it. This approach differs from the contemporary population analysis or historical demography, which use available data directly to analyse contemporary or past populations.

I have been working on this booklet, slowly, since 2017, and in that time, the research environment has improved tremendously. I am deeply grateful for the digital archives provided by public sources like the National Diet Library and the National Archives of Japan. Most historical books and materials are now available in a digital format and can be downloaded by registered researchers through internet. This has been an enormous contribution to historical studies, allowing researchers to conduct source-finding efforts from a computer, rather than visiting libraries. The time spent copying or photographing materials has been substantially reduced, simply by downloading PDF. Although these materials are in Japanese, the electronic format facilitates OCR and machine translation, which will open the way for analysis by scholars worldwide. I see my role as simply introducing this wealth of information by providing URLs, thereby paving the way for further research.

I extend my gratitude to Dr. Aya Homei for her insightful comments on my manuscript and for expanding my knowledge in the field. I also thank the Editorial Committee, launched by Prof. Toshihiko Hara and now chaired by Dr. Toru Suzuki, for this opportunity. I am also very grateful to Mr. Yutaka Hirachi of Springer for his tireless patience with my slow writing.

This work was supported by various research grants from 2017 to 2025, particularly the Population and Social Security Research Archive Formation Project of the National Institute of Population and Social Security Research (IPSS) and the Research Grants of Ministry of Health, Labour and Welfare (H29-Seisaku-Shitei-003, 20AA2007, 23AA2005, 23AB0201).

Tokyo, Japan   Reiko Hayashi

Contents

1	**Fertility: Contradicting Perceptions and Ambivalent Policies**	1
	1.1 Introduction	1
	1.2 The Crime of Abortion and Birth Control	3
	1.3 The Commission for the Investigation of Problems of Population and Food	4
	1.4 The Outline for the Establishment of Population Policy	6
	1.5 The Most Drastic Fertility Changes in Japanese History: Post-War Baby Boom, Eugenic Protection Act, and Family Planning	7
	1.6 Hinoe-Uma: The Year of Fire-Horse	9
	1.7 The Period of Indifference (1972–1989)	12
	1.8 Measures to Tackle Low Fertility	13
	1.9 Continuously Evolving Policies with Uncertain Results	14
	References	15
2	**Mortality: Epidemiological Transition and Health System Development**	19
	2.1 Introduction	19
	2.2 The Creation of Health Administration and the Cause of Death Statistics in Early Meiji	20
	2.3 The Fights Against Infectious Diseases	23
	2.3.1 Cholera	23
	2.3.2 Tuberculosis	24
	2.3.3 Epidemic Cold (Spanish Influenza)	27
	2.4 1920: The Onset of Mortality Decline	28
	2.4.1 The Quality of Statistics for Infant Mortality	28
	2.4.2 Infant Mortality by Cause of Death and by Prefecture	30
	2.4.3 The Determinants of Infant Mortality Decline	30

	2.5	The Path to Universality: Health Insurance Expansion	33
	2.6	Onset of Ageing	36
	2.7	Reforms, Consumption Tax and the Welfare Bubble	39
	2.8	Toward the Limit of Human Longevity	40
		References	41
3	**Migration: The Most Effective Population Policy?**		45
	3.1	Introduction	45
	3.2	Pre-Modern Mobility: In the Edo Period	47
	3.3	Meiji Restoration as the Liberalization of Mobility	49
	3.4	Shifting Destinations Amid an Evolving International Climate: Emigration to Ease the Population Pressure	51
	3.5	Peopling the Greater East Asia Co-Prosperity Sphere	54
	3.6	Repatriation	56
	3.7	The Creation of New Migration Statistics and the Surge of Post-War Internal Migration	58
	3.8	Labour Placement: From Order, Assistance to Choice	59
	3.9	From a Sending to a Receiving Nation: The Transformation of International Mobility	60
	3.10	Migration Policies in an Era of Population Decline	63
		References	64

Chapter 1
Fertility: Contradicting Perceptions and Ambivalent Policies

1.1 Introduction

1925 was the first year that the number of births by mother's age became available, enabling the calculation of the total fertility rate. In less than a century, Japan's total fertility rate dropped from 5.11 in 1925 to 1.26 in 2005. Before that, more straightforward fertility measurements, such as the number of births or the crude birth rate, were available. The number of births was on the rise until the start of World War II. The crude birth rate, defined as the number of births divided by the total population, rose from 1872 to 1900, stagnated, and then declined (Fig. 1.1). The drop for the decade from 1947 to 1957 was sharp, but throughout the period, the trend was never uniform.

From the start of the Meiji period, the change in fertility can be roughly divided into four phases. In the first phase, from 1872 to 1919, fertility gradually increased, with some exceptions. In the following phase, from 1920 to 1945, fertility decreased gradually until it dropped suddenly in 1938, then turned upward. In the third phase, from 1946 to 1973, a sharp fertility decline was followed by a period of stagnation. In the fourth phase, starting from 1974, a steady fertility decline continued until 2005, then turned to an upward trend, followed by a downturn again.

The Family Register Act was promulgated in 1871 and implemented in February 1872 [1]. The number of births that year was 569,034, 30% less than the following year. This was due to the omission of January births, and also, it also included births only until December 2nd 1872 was a historical year in which the calendar system was switched from the Tenpo calendar, the official calendar in use since 1844, to the Gregorian calendar. The year 1872 ended on December 2nd, and January 1st 1873, started the following day. From the next year, in 1873, the number of births and the crude birth rate rose.

Many later demographers contended that these early statistics on the number of births were afflicted by under-registration and assumed birth rates were much higher

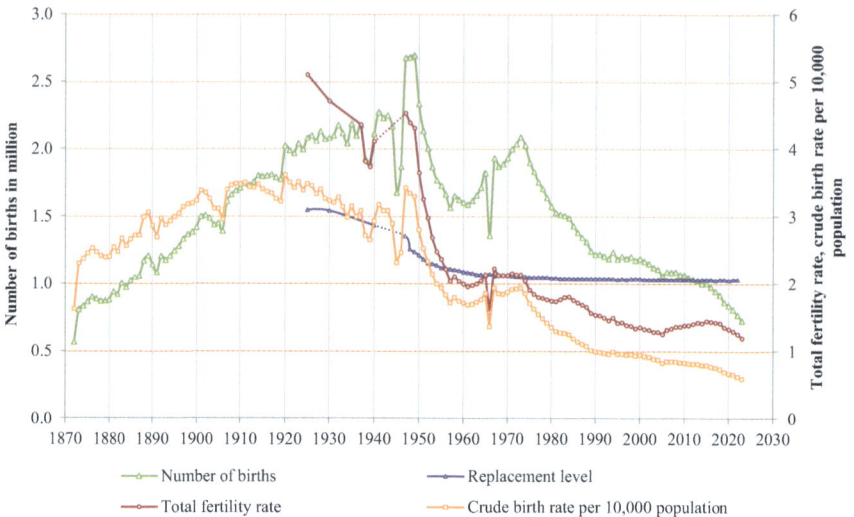

Fig. 1.1 Trend of fertility in Japan. Sources: Number of births, crude birth rate and total fertility rate by Vital Statistics (from 1872 to 1943 by Statistical Bureau, from 1944 onward by Ministry of Health (Labour) and Welfare), replacement level by Population Statistics of Japan (National Institute of Population and Social Security Research)

(Fig. 1.2). However, they agree that the discrepancy was only until 1880s. The revision of the Family Register Act in 1886 improved the quality of registration and the backward projection proved the number of birth since 1890 was at a good level of accuracy [6]. By that Act, births were to be reported within 10 days, and burial should be made after death registration. De-registration was also put in place. Statistics on the monthly number of births and population tables by single age became available. Along with the Cemetery and Burial Regulations stipulated two years earlier, in 1884, stillbirth statistics began to be compiled. This was the time when population statistics based on the family register were modernized and rationalized.

The Bureau of Family Register compiled and published statistics on births based on registration until the Cabinet Statistics Bureau, established in 1898, replaced it. With another revision of the Family Register Act in 1898, the number of births became an element of the Vital Statistics.

In these earlier days, although statistics on the number of births were available, it was difficult to find contemporary arguments on fertility trends and analyses, let alone policies on fertility. During this period, the term "population" was not recognized as it is today [7]. Simply, the growth in the number of people was considered a country's wealth [8]. The fact that there were multiple contradicting estimations suggests that there was, and had been, no consensus on the fertility trend, especially up to 1890. If there was no perception of particular population behavior, then there was no chance that a population policy could exist. Probably the only

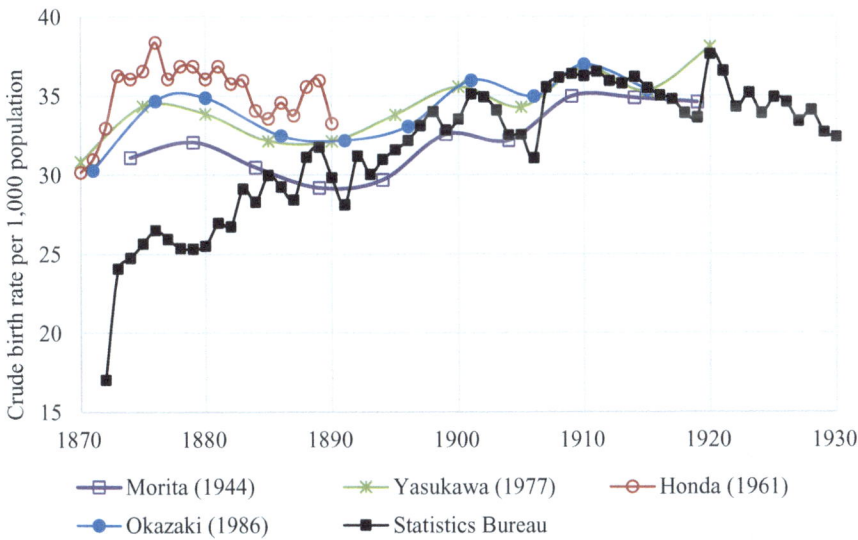

Fig. 1.2 Various estimates of crude birth rates by later demographers [2–5]

fertility-related governmental action was to reduce the "bad habit" of controlling the number of children, namely Mabiki or infanticide [9].

1.2 The Crime of Abortion and Birth Control

In Japan, accurate knowledge of human reproduction has been shared since the beginning of the Meiji era [10]. Unlike Catholic Western countries where sexual abstinence was the norm and abortion was prohibited, various contraceptive methods were known in Japan, though not necessarily effective, and even abortion and infanticide were tolerated as a necessity. Although the governmental notice on the abortion interdiction was released in 1868, its effectiveness was doubted [11]. In addition, the Shinko Ritsuryo, the first penal code elaborated by the government in 1870, did not mention abortion and infanticide. The crime of abortion emerged in the 1882 penal code, which was instructed and drafted by Gustave Émile Boissonade, the French jurist whom the Meiji government invited to modernize Japanese laws. The texts resembled Article 317 of the French Penal Code of 1810. Some later scholars argue that the crime of abortion stipulated in the Penal Code since 1882 was to control women's fertility [12], but it is mostly agreed that this crime of abortion was not intended as a pronatalistic policy but simply to show the outside world that Japan was a "modernized country". Abortion control was utterly ineffective, at least at the beginning [13–15]. Additionally, among contemporary statisticians, the crime of abortion was classified in the section of public order, not population or fertility. There was only one article on abortion published in the *Journal of the Statistical Society of Tokyo*, the very first journal on statistics that also provided information on the perception of statisticians from 1880 onwards. That article, published in 1881, listed the number of abortions by prefecture, along with the number of robberies and gambling [16]. Abortion was a statistical subject in the domain of crime, not population control.

However, around the turn of the twentieth century, abortion and fertility control started to be recognized by demographers and statisticians. The earlier introduction of Malthusian theory in 1870 raised awareness of population increase, but it was not necessarily linked with the role of fertility control [17]. Then, Neo-Malthusianism arrived and influenced the Japanese people. A prominent Japanese pharmaceutical businessman observed Annie Besant's activities during his stay in England and was enlightened. He advocated birth control for social improvement [18]. Statisticians also began to recognize that fertility change was not merely caused by natural and biological factors but also by social factors [19]. However, socialists criticized neo-Malthusian thinking, and the idea of fertility control did not expand during this period [20].

1.3 The Commission for the Investigation of Problems of Population and Food

The rice riot of 1918 reminded people of the population problem. Rice production in 1917 fell to 54.6 million Koku (8.2 million tons), a 7% decline from the previous year [21]. As a result, prices soared, and panic spread among the people. It has been assumed that "there was no population problem in the Meiji era" [9], and the rice riot in 1918, the seventh year of the Taisho era, marked the first emergence of the population problem. There had been many rice riots before 1918, as well as several years when rice production plunged compared to the preceding year. What was new in 1918 was that people began to link rice shortage and riots to overpopulation. This shift in perception coincided with broader developments: the launch of the vital statistics in 1899, the long preparations for the first population census, finally conducted in 1920, the visit of Margaret Sanger to Japan in 1922, and the first full publication of Malthus's *An Essay on the Principle of Population* into Japanese in 1924 [22]. The rice riot occurred when people were beginning to recognize there were such things as "population" and "population problem". It was not merely the lives of individuals but rather the population, the aggregate number of human beings, was now seen as something that could cause problems and potentially disrupt the functioning of society.

Under such circumstances, the Commission for the Investigation of Problems of Population and Food (CIPPF) was established in 1927. It was the first governmental body to deal with the population problem, stipulated by Imperial Edict and presided over by the Prime Minister. The population problem identified by the Commission was that of overpopulation. The Commission's main objective was to balance food and population and alleviate the nation's worries. The Commission was divided into two sections, food and population, and within four years, seven reports on food and six reports on population were delivered. The population six reports were: "International and Internal Migration"(1927), "Measures for Labour Supply-demand Control"(1927), "Population Countermeasures in Regions Outside of Japan Proper"(1928), "Various Measures for Population Control"(1929), "Report for Productivity Increase"(1929), and "Measures concerning Distribution and Consumption"(1930) [23]. Although fertility control is an essential aspect of population control, there was no separate report focusing on fertility. Those issues were included in the report "Various Measures for the Population Control".

At the beginning of the Commission, a document was submitted that described in detail the practices of birth control inside and outside Japan [24]. This document, elaborated by Shigenobu Masuda, who later became a researcher at the Foundation-Institute for Research of Population Problems (FIRPP), was classified as a "secret" document by CIPPF. However, it was full of useful information expressed from a neutral point of view. It ended with a policy suggestion that birth control-related businesses should be controlled appropriately, banning harmful ones, permitting harmless ones, and prioritizing the counseling centers for poor people. It is not clear how the Commission used this document. Still, after its submission in February 1928, the meetings on eugenics and birth control became the most heavily debated matter in the Commission. Toru Nagai and Hisomu Nagai submitted each a proposal to prepare the report. The former emphasized the importance of birth control, and the latter emphasized racial hygiene and proposed a law to allow "rational" contraception and abortion. This point-the promotion of birth control-seemed to be discussed fiercely in the following three meetings and was finally deleted at the fourth meeting held on February 6th, 1929 [25]. The resulting report only included control over the illegal sale of contraceptives and a promotion of research on eugenic policies. It was a time when birth control and abortion were considered immoral and criminal by the general public and, more importantly, by policymakers at large.

In addition to the six reports, the Commission adopted two decisions in 1930: to establish a population research institute and a ministry in charge of social affairs. For the former, the Foundation-Institute for Research of Population Problems (FIRPP), a non-profit private organization, was established three years later in 1933 to pave the way for establishing a proper governmental organization [26]. For the latter, the Ministry of Social Affairs never came into existence. Instead, the Ministry of Health and Welfare was established in 1938 to compromise between different motivations for social affairs and the need for strong and hygienic human resources ready to fight in the war [9, 27, 28].

1.4 The Outline for the Establishment of Population Policy

At the time when CIPPF reported the need to create a ministry in charge of social affairs and a research institute on population, the main objective was to ease overpopulation. Together with Socialism, Communism, and Marx's theory, scholars in the field of economics discussed the theory of population. However, in the following decade, the policy direction diversified, and the need to promote the population from a health perspective grew. The year 1937 was a turning point for the population problem. In July of that year, the government organized a committee, assembling the FIRPP and the Statistical Society of Tokyo members to participate in the fourth International Conference on Population organized by the International Population Union in Paris [29]. After the Conference, the first meeting of the National Council on Population Problems was held in November of the same year, where 73 presentations were delivered, gathering around 300 academicians and politicians interested in population problems. At the opening session, Prime Minister Fumimaro Konoe made a congratulatory remark. Also, at this meeting, a petition to establish a national research institute on population was unanimously adopted [30].

Finally, in 1939, the Institute of Population Problems (IPP) was created. One of the first reports published by IPP was on the sudden drop in fertility in 1938 and 1939 [31]. That was a concern well shared within the government. The vital statistics report suddenly added the table of births disaggregated by the mothers' age from 1937 to "enable the research on the fecundity" [33]. The Sino-Japanese War brought many young Japanese men to China, and a scientific explanation was necessary. The IPP report concluded the fertility drop was due to the drafting of young married men as soldiers to China and also the consequent internal migration of male workers from rural to urban areas. This observation affected the drafting method in the following years. Younger, unmarried men were recruited, and an alternate return system was taken, which could have contributed to the fertility increase from 1940 [34]. Another important work of IPP at this early stage was population projection. The first version of the projection was published in the IPP's *Journal of Population Problems* [35], and these projections became the basic theoretical foundation for the development of the Greater East Asia Co-Prosperity Sphere. In line with this basic population research, IPP also elaborated the first version of the Outline of Population Policy in 1940 [32], which was extensively modified by the Planning Board and decided by the Cabinet as the Outline for the Establishment of Population Policy in 1941 [36].

The Outline is the first and only "Population Policy (*jinkō seisaku*)" in the history of Japan, in the way that the term was used as the title. It aimed for rapid and eternal population development in the Greater East Asia Co-Prosperity Sphere, and the population would reach 100 million in 1960. For that purpose, measures on fertility, mortality, and population distribution were stipulated. For fertility, women's employment after 20 years old was recommended to be abandoned, marriage should be hastened by three years, and each couple should have five children. A family allowance system, a national fund, and tax exemption for large families were proposed. Contraception and induced abortion were prohibited. For mortality, the primary objectives were infant mortality reduction and tuberculosis prevention, and 35% mortality reduction was set as the target. For that purpose, the strengthening of the network of health systems composed of health centers and the expansion of the health insurance system were planned. Based on the rapid urbanization at the time, population decentralization and rural development were promoted. The underlying philosophy was the importance of family and race, and individualism was negated.

The Outline has been the icon of wartime totalitarian policy, and it was the source causing the taboo on any pronatalistic policy in the post-war era. The term "population policy" was avoided as it reminded people of the Outline. However, the effectiveness of the policy has never been proven. To begin with, Mr. Koizumi, who took office as the Minister of Health and Welfare six months after the adoption of the Outline in January 1941, almost neglected the part on fertility policies but prioritized mortality policies [28]. From the year the Outline was adopted until the end of WWII, the crude birth rates decreased (Fig. 1.1), and it is difficult to find any budgetary evidence of the implementation of measures, apart from the improvement of National Health Insurance coverage.

1.5 The Most Drastic Fertility Changes in Japanese History: Post-War Baby Boom, Eugenic Protection Act, and Family Planning

The post-WWII baby boom started right after the end of the war and lasted for many years in Western countries, but in Japan, it started a bit later and for a shorter duration. The war ended in August 1945, and a significant increase in births occurred not in 1946 but in 1947. As many as 7 million overseas Japanese, composed of outmigrants and soldiers, returned after the war, and it needed a while to resume normal life. But also, the monthly births decreased significantly in December 1946 and increased just as much in January 1947, which delayed the onset of the baby boom when annual figures are concerned. This phenomenon, even apparent in the first statistics on monthly births of the year 1880 [37], is due to the people's preference to make the child one year younger, taking into account the Japanese age calculation method at the time [38–40]. Many births in the last months of 1946 could have been registered in the early months of 1947. So, the onset of the baby boom in post-war Japan could have started in 1946, if correctly registered, and was not so different from those in other post-war countries.

However, what makes the Japanese post-WWII baby boom unique is its brevity of duration. After the sharp rise in fertility in 1947, the high level continued for only three years until 1949 (Fig. 1.1). This was due to policy intervention. In 1948, the Eugenic Protection Law was enacted and opened the way for induced abortion. The law was drafted and proposed by members of parliament, including Mrs. Shizue Kato, one of the first 39 women parliamentarians and an eminent advocate of birth control who invited Mrs. Margaret Sanger to Japan back in 1922. The law, together with the revisions in 1949 and 1952, made induced abortion available even for economic reasons. The number of abortions soared, and fertility declined rapidly. In terms of the total fertility rate, in only nine years from 1947 to 1956, it declined from 4.54 to 2.22, below the replacement level of 2.24 of that year (Fig. 1.1). According to official statistics, the number of induced abortions was 1.17 million in 1955, the highest in history, comprising one-third of total pregnancies, but some argue that the proportion is even higher as many abortions were not registered [41]. Naturally, there were concerns over this rapid expansion of abortion [42], but it was the time when pre-war conservative political figures were purged from public function, and the Allied Powers let the Japanese decide, avoiding interference in the matter of population control [8, 43].

Governmental actors on population issues were organized at this time. The Advisory Council on Population Problems (ACPP) was created first in 1949 by the Cabinet and renewed in 1953 within the Ministry of Health and Welfare. Also, the Foundation-Institute for Research of Population Problems (FIRPP) regained its activity in 1951. Together with the Institute of Population Problems (IPP), the three organizations on population problems formed a framework to elaborate the population related policies based on data and research [44].

On the other hand, family planning policies and activities were strengthened. The FIRPP compiled a list of opinions on birth control expressed in the media and by eminent personage in 1946 [45]. The IPP conducted the first birth control survey in 1947 [46]. Mainichi Newspapers established the Population Problems Research Council in 1949 and conducted opinion polls on family planning [47]. In 1951, a cabinet decision was made to promote family planning to reduce the harmful effect on the body of the mother, and a governmental budget of 21.2 million yen was allocated in 1952, which doubled to 58.7 million yen in 1955 [41]. The Population Association of Japan, created in 1949, started receiving support in 1951 from an American philanthropist, C.J. Gamble [48], who was committed to promoting family planning worldwide [49]. The Japan Family Planning Association was created in 1954, and it disseminated and promoted information and methods of family planning. Activities of all these organizations focused mainly on reducing fertility and easing population pressure.

During this period, when lowering fertility was the main target, there was no fear of low fertility. The population projection conducted by IPP in 1955 and 1957 assumed the total fertility rate to decrease until it would hit a bottom of 1.6 and then stabilize [50, 51]. No concern was found in the written materials by the researchers involved with these projections that the 1.6 total fertility rate was too low, regardless of the fact that the rate of 1.6 is much lower than the replacement level, which would lead to population extinction in the long run. Some IPP researchers started to express concerns about the low level of fertility [52–54], and the ACPP report of 1969 revealed that the fertility level could be one of the lowest among countries in the world. It emphasized the need for social development in harmony with economic development to attain fertility recovery [55]. But after all, the overall fertility level remained around the replacement level until the 1970s, although there was a significant fluctuation of fertility, notably in 1966, the year of Fire-horse, and the population issues focus shifted to population ageing.

1.6 Hinoe-Uma: The Year of Fire-Horse

Hinoe-uma, the year of Fire-horse, is a year assigned every 60 years by the traditional Ganzhi calendar, with the most recent one being 1966. According to the traditional Chinese year-counting system, which has also been used in Korea and Vietnam, the 10 signs (stems) and 12 animals (branches) are combined, and based on the arrangement rule, 60 combinations are assigned each year. Fire-horse year is common in all countries, but it is only in Japan that the year is linked to a superstition that a girl born this year would eventually "eat up" her husband. The belief started in the early Edo period based on an episode of an arsonist woman who was allegedly born in 1666, the year of Fire-horse [56]. Following this event, the next Fire-horse year of 1726 saw an increase in women's deaths due to the inappropriate usage of strong laxatives to avoid pregnancy. Similar anecdotes were found in 1786, and the birth shortage was also confirmed using modern statistics for those born in the year 1846 [57].

The next *Hinoe-uma* in 1906 was noticed by contemporary statisticians with the low birth sex ratio in that year, and it was attributed to the timing of girls' birth registration [58]. However, the fertility decline in this year was mixed with the effect of the Russo-Japanese War, which continued until September 1905, and the fertility single-year drop was not as noticeable compared to 1966.

The most recent year of Fire-horse, in 1966, saw the most drastic fertility decline expressed in modern statistics. The number of births declined by 25%, from 1,823,697 births in 1965 to 1,360,974 in 1966, a deficit of 462,723 births. The following year, 1967, showed a notable rebound to 1,935,647 births. Many contemporary administrators, statisticians, and demographers were surprised such a thing would happen in modern society. Surveys, observations, and analyses have been conducted since then. The decrease in births was partly due to shifting birth registration from January 1966 to December 1965 or December 1966 to January 1967, as in the previous Fire-horse year of 1906. However, it explained only 2% of the decrease. The decrease was not caused by abortion as its number even declined in 1966 compared to 1965. So, the 462,723 births decrease was realized by enforcing family planning [59]. According to the survey rapidly conducted in September 1966 by the Ministry of Health and Welfare, almost all (98%) married women questioned knew about the year of the Fire-horse. However, only 4% avoided births due to Fire-horse. 24% of women ambiguously answered they did not expect a baby for a while. These women could have contributed to the decline of births. The decision to have babies is not a strong determination but a floating mind affected by the mood in society. An increased usage of contraception was observed compared to normal years. They probably did not have a strong wish to avoid the Fire-horse baby, but somehow increased the usage of family planning and let the pregnancy happen in the near future. Or, the Fire-horse year superstition was a good excuse for women to avoid pregnancy, which they were supposed to have but did not wish.

The sex ratio at birth in 1966 was 107.6, slightly higher than the standard sex ratio of 105. As the superstition concerns only baby girls, the high sex ratio, more baby boys than baby girls, is understandable. However, how was this attained? At the time, there was no ultrasound technology to conduct sex-selective abortions. Media articles added information on how to give birth to desired sex (e.g. [60]), but the effectiveness was not known. A simple solution was shifting the registration. Baby girls born in January 1966 were registered in December 1965, and those born in December 1966 were registered in January 1967. The irregular sex ratio at birth was only found in those months (Fig. 1.3), and these monthly distortions were reflected in the annual figures.

What has been noticed but not well examined until now was the increase in suicide and single-mother households in 1966 [61]. Compared to 1965, there were 606 (4%) more suicides in 1966. Women were more affected (Fig. 1.4), and one-fifth of the increase was attributed to young (15–19 years old) women. Anecdotally, suicides among women born in 1906 are known when they faced difficulties in getting married. The number of marriages was suppressed in 1965 and 1966 (Fig. 1.5), possibly due to couples who feared having a baby right after the marriage. The marriage postponement or cancellation could have put some young women in mental turmoil. As for the increase in single-mother households, the dispute among married couples on reproductive choices for the year of Fire-horse could have caused divorce and an increased number of single-mother households. The number of divorces during these years has not shown a similar increase-decrease trend, but it has increased monotonously since then. The divorce-marriage proportion hit its bottom then started to increase from 1965. From a longer-term point of view, the Fire-horse year of 1966 was one of Japan's turning points in marriage and divorce.

Fire-horse superstition in 1966 affected some who believed a baby girl born in that year was a misfortune. But probably, for most people, it worked as a revelator

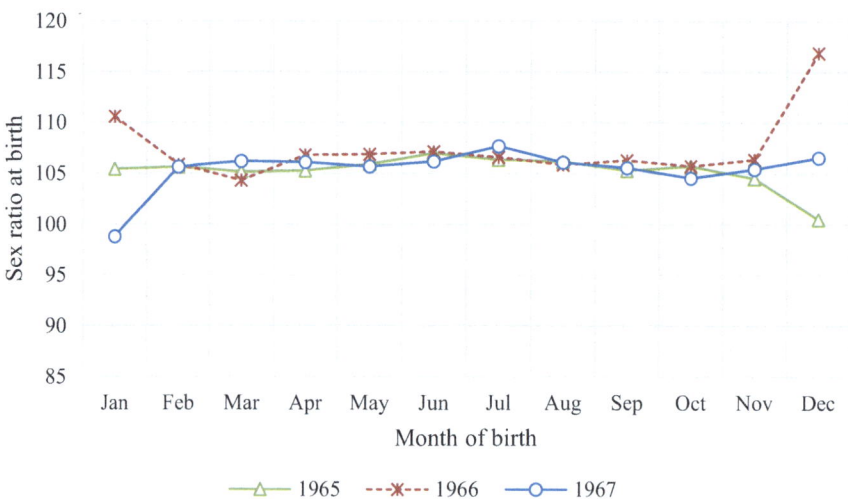

Fig. 1.3 Sex ratio at birth around the year of Fire-horse. Source: Vital Statistics (Ministry of Health and Welfare)

1.6 Hinoe-Uma: The Year of Fire-Horse

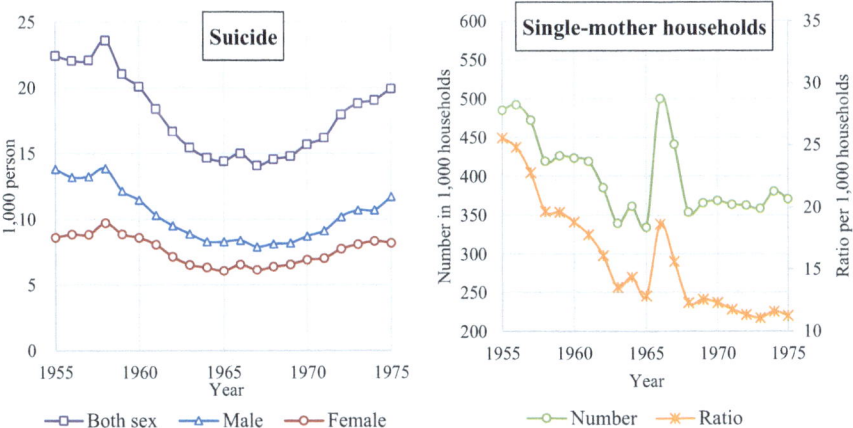

Fig. 1.4 Changes in suicide and single-mother households in the year of Fire-horse. Note: The number of single-mother households in 1966 was surveyed in the Ministry of Health and Welfare Survey on Comprehensive Life Survey, but the estimates of the household number were later adjusted. The above figure uses the adjusted household number. Sources: suicide by Vital Statistics, single-mother households by Comprehensive Survey of Welfare Administration (both by Ministry of Health and Welfare)

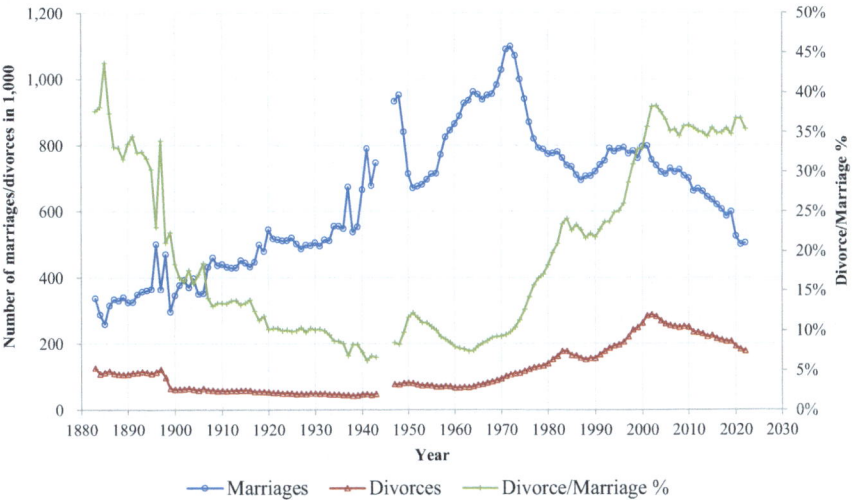

Fig. 1.5 Changes in marriage and divorce. Sources: Statistical Yearbook (Statistics Bureau) from 1883 to 1897, Vital Statistics (Statistics Bureau and Ministry of Health (Labour) and Welfare) for 1898 to 2022

to expose hidden wishes and needs regarding people's choices on marriage and having children. The Fire-horse birth deficit and the onset of divorce increase were triggered by the Fire-horse, which were otherwise sealed off by cultural norms. It was a one-year event, but the starter of the consecutive trend in the following years. The total fertility rate kept on decreasing until it went below that of the Fire-horse year in 1989, and divorces kept on increasing (Fig. 1.5).

1.7 The Period of Indifference (1972–1989)

The stagnant fertility never lasted. In the "population year" of 1974, when the Japan Population Conference was held in Tokyo and the World Population Conference was held in Bucharest, Romania, the direction was to lower fertility. The theme of the Tokyo Conference was "towards a stable population", and the level of fertility, as well as the increasing population due to population momentum, were considered not optimal given the limited resources available [62]. A similar message was also delivered at the United Nations World Population Conference held in Bucharest, Romania, one month after the Japan Population Conference. Those engaged in the Japan Population Conference seemed to follow what the UN advocated, regardless of their own country's intrinsic population trend. In fact, the analysis made in the Japan Population Whitepaper prepared for the Conference was full of descriptions that suggested possible threats to maintaining the level of the Japanese population. Still, the declaration made at the Conference was in the opposite direction. The conference's declaration stated, "in 50 years, it is certain the Japanese population will reach 140 million..." It then called for "measures to contain the population increase" and advocated for a "national consensus on having no more than two children". The strange gap might be due to the taboo against shifting the population policy toward raising fertility. It might also be because notable population specialists passed away just before the Conference. Minoru Tachi, then the Director-General of IPP, suddenly passed away in 1972. Toru Nagai, then the Chairman of the Foundation-Institute for Research of Population Problems, died in 1973. Yoshio Koya, then the Chairman of JOICFP (Japanese Organization for International Cooperation for Family Planning) followed them in 1974, just before the conference started. It is possible that the conference failed to reflect a proper understanding of the population problem, without those brains.

As recommended and endorsed by the Conference, Japanese fertility started to plunge further down from the total fertility rate of 2.05 in 1974 to 1.57 in 1989. Despite Japan's consistently lowering fertility rates—with only four countries (Germany, Luxembourg, Italy, and Austria) having a continuously lower rate during the 1980s [63]—there was a notable policy vacuum until the establishment of the "intra-governmental network meeting to create the environment to foster children" in 1990. Several factors contributed to this lack of responsive policy during this period. First, the fear of low fertility was not widely shared in society. As evidenced by the 1974 conference, the focus was still on lowering fertility, not increasing it. Even the specialists on this matter considered that fertility would soon rebound. The three population projections conducted in 1976, 1981 and 1986, consistently assumed a fertility increase from the projection year. This assumption was based on survey data indicating that the completed, the ideal, and intended number of children, remained stable at approximately two per married couple. Indeed, marital fertility did not decline during this time, but the nuptiality rate, particularly among younger age groups (e.g., ages 20–24), decreased significantly. This trend was optimistically interpreted as a delay in marriage rather than a permanent rejection of it, leading to the expectation of an eventual fertility rebound. However, the lifetime celibacy rate, the proportion of individuals who never married by age 50, steadily increased, and fertility continued to decline.

Another reason for the policy delay can be attributed to the strong societal taboo against pronatalist policies, a reaction to the coercive population policies of the wartime era. While family policies, such as child allowances (cash payment to the families with children), or provision of childcare facilities were implemented, they were framed as mother-and-child welfare, rather than encouraging to have more children. For example, during the discussion to create the child allowance system in 1972, the notion of the pronatalistic effect of child allowance was carefully avoided [64]. Similarly, childcare facilities were established by municipal governments under the Child Welfare Act of 1947 to parents unable to care for their children, and these measures were considered social welfare, not to facilitate women's employment. Furthermore, paid childcare leave for working mothers was not institutionalized until 1992. In short, during a time when the ordinary life course of women was considered to be getting married, quitting jobs, marrying, and raising children, low fertility concerns and family policies were not yet conceptually linked, leaving a limited ragne of policy options to address the low fertility.

1.8 Measures to Tackle Low Fertility

In 1990, the so-called "1.57 shock" hit the public. The total fertility rate of the previous year (1989) was recorded at 1.57, below the level of the 1966 Fire-horse year (1.58). This demographic alarm bell sounded at a time when the status of women in Japan was undergoing significant change. In 1985, Japan ratified the Convention on the Elimination of All Forms of Discrimination Against Women and enacted the Equal Employment Opportunity Law, theoretically allowing women to work like men. However cultural norms lagged behind legal reforms. Despite these new opportunities, traditional gender roles persisted. From 1980 to 1984, one in four marriages were still arranged [65], and it was customary for many women to leave their jobs upon marriage and childbirth. Although they became able to get equal jobs as men, once they got married, household chores and child-rearing fell almost entirely on them. The cost of hiring a nanny often exceeded a mother's entire salary. The "1.57 shock" was a trigger that made people think of these contradictions.

In response, the government established the "intra-governmental network meeting to create the environment to foster children" in August 1990 [66], which led to a series of consecutive governmental plans to tackle low fertility (Fig. 1.6). These plans broadened the scope of policy areas over time. While these measures did not

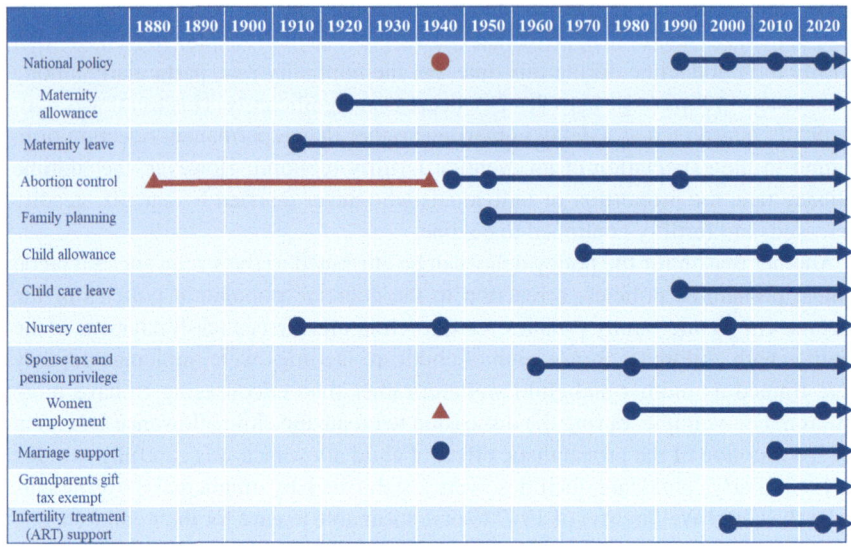

Fig. 1.6 Fertility-related policies in Japan (1880–2020)

provide a quick fix, they at least increased public awareness of the issue. The total fertility rate bottomed out at 1.29 in 2005 before rising temporarily, only to begin another decline in 2015.

1.9 Continuously Evolving Policies with Uncertain Results

In June 2016, the Cabinet adopted Japan's Plan for Dynamic Engagement of All Citizens. For the first time since the end of WWII, a numerical fertility target was incorporated into governmental policy. The "Desirable Fertility", set at 1.8 at the time, was defined as the possible total fertility rate if all the individuals fulfilled their wish to marry and have their desired number of children, based on data from the National Fertility Survey conducted by IPSS. Before this Cabinet decision, the term was discussed in various governmental and non-governmental meetings and commissions. The Minister in charge at the time, Shigeru Ishiba, clearly stated that the government should not interfere with personal fertility decisions but rather remove barriers preventing people from marrying and having their desired number of children [67].

Had this discussion occurred ten years earlier, the "desirable fertility" target likely would not have been accepted as it was. Setting a numerical target for fertility had been taboo, as it evoked memories of the coercive governmental control over family life stipulated in the 1941 Outline for the Establishment of Population Policy. As recently as 2013, civil society strongly opposed the governmental plan to distribute the Women's Handbook, viewing it as a tool for governmental control. The

Women's Handbook, analogous to the Maternal and Child Health (MCH) Handbook, was intended to better manage women's health by recording reproductive health-related events. So, why was the Women's Handbook rejected while the Desirable Fertility 1.8 was accepted? One reason could be that the population decline had steadily progressed over the years and gained wider recognition. People, as well as national and municipal governments, began to feel the threat and believed substantial measures were necessary. Another reason might be the declining influence of organized civil society focused on women's empowerment. Notable figures who advocated for women's status throughout the Showa era were ageing. The younger generation benefited from gender equality and equal employment opportunities with men, making them less provocative, and more focused on their business carriers. Similarly, as labour unions lose members and become more conformist, women's activists might now be less rebellious than before.

Throughout the history of modern Japan since the Meiji Era, the perception of fertility and the policy responses to it have taken many forms (Fig. 1.6). Fertility was well perceived, and action was taken in the 1940s and after the 1990s, albeit with different approaches. It was neglected or avoided from discussion before 1920 or in the 1970s and 1980s. Fertility policies change direction according to the specific needs of each period; the importance lies in how people perceive the situation and whether they have responded, not solely in evaluating the policy's impact and efficiency.

References

1. Cabinet Official Gazette Bureau. 1888. Meiji 4 Hourei Zensho (Compendium of laws). https://dl.ndl.go.jp/pid/787951
2. Morita, Yūzo. 1944. *Jinkō Zōka no Bunseki (Analysis on population increase)*. Nippon Hyoron Sha. https://dl.ndl.go.jp/pid/1459595.
3. Yasukawa, Masaaki. 1977. *Meiji Taishō nenkan no jinkō suikei to jinkō dōtai. Jinkō no Keizaigaku (Economics of population)*, 149–189. Shunjusha.
4. Honda, Tatsuo. 1961. *Review of Japan's vital statistics since around the Meiji restoration*. Annual report of the Institute of Population Problems, Institute of Population Problems, Ministry of Health and Welfare, Japan.
5. Okazaki, Yoichi. 1986. Population of Japan in the Meiji-Taisho era - Re-estimation. *Journal of Population Problems (Jinkō Mondai Kenkyū)* 178:1–17. https://www.ipss.go.jp/syoushika/bunken/data/pdf/14167501.pdf.
6. Takase, Masato. 1991. Population, birth and death in Japan for the period 1890-1920. *Jinkōgaku Kenkyū (The Journal of Population Studies)* 14:21–34. https://doi.org/10.24454/jps.14.0_21.
7. FIRPP: Foundation Institute for Research on Population Problems. 1983. *Jinko Mondai Kenkyukai 50 nen Ryakushi (Brief history of 50 years of Foundation Institute for research on population problems)*. Jinko Joho, Fiscal Year of Showa 57.
8. Taeuber, Irene B. 1958. *The population of Japan*. Princeton University Press.
9. Ministry of Health and Welfare Editorial Committee for 50 years History of Ministry of Health and Welfare. 1988. *Kōseishō 50 nen shi (50 years history of Ministry of Health and Welfare)*. Foundation-Institute for Research of Health and Welfare Problems.

10. Hirano, Sukezo. 1880. Methods of conception and contraception control. Chieno-Ko No. 45 Appendix, Usagiya Makoto. In *Sei to Seishoku no Jinken Mondai Shiryō Shūsei*, vol. 1. Fuji Shuppan.
11. Koizumi, Eiichi. 1934. *Dataizai Kenkyū (Research on crime of abortion)*. Ganshodo Shoten.
12. Fujime, Yuki. 1998. *Sei no Rekishi Gaku - Kōshō Seido· • Dataizai Taisei kara Baishun Bōshi Hō • Yūsei Hogo Hō Taisei e (History of sex - From the regime of licensed prostitution and crime on abortion to the regime of anti-prostitution law and eugenics protection law)*. Fuji Shuppan.
13. Drixler, Fabian. 2013. *Mabiki - Infancicide and population growth in eastern Japan, 1660–1950*. University of California Press.
14. Ishizaki, Shoko. 2015. *Kingendai Nihon no Kazoku Keisei to Shusshoji su (Family formation and number of children in modern Japan)*. Akashi Shoten.
15. Iwata, Shigenori. 2009. *Inochi o meguru kindaishi*. Rekishi Bunka Library 271, Kikukawa Kobundo.
16. Statistical Society of Tokyo. 1881. Kaku Fu Ken Datai (Abortion by prefecture). *Journal of the Statistical Society of Tokyo* 4:209–211.
17. Oshima, Sadamasu. 1877. Marusasu Jinko Ron Yoryaku (The overview of the theory of population of Malthus). https://dl.ndl.go.jp/pid/800901
18. Oguri, Sadao, and Kanichiro Garai. 1903. *Shakai kairyō jitsuron (The real theory of social betterment)*. Reprinted in: *Sei to Seishoku no Jinken Mondai Shiryō Shūsei*. Vol. 1. Fuji Shuppan.
19. Kawai, Toshiyasu. 1907. Wagahō no shusshōryoku (Fertility in Japan). *Journal of the Statistical Society of Tokyo* 318:395–397.
20. Ota, Tenrei. 1976. *Nihon sanji chōsetsu hyakunenshi - Meiji Taishō Shōwa shoki made (100 years history of birth control in Japan - From the Meiji, Taisho and early Showa periods)*. Shuppan Kagaku Sōgō Kenkyusho.
21. Statistics Bureau. 1918. The 37th statistical yearbook of Japanese empire. https://dl.ndl.go.jp/info:ndljp/pid/974425
22. Takano, Iwazaburo, and Bei Ouchi, translation. 1924. *Jinkō no genri ni kansuru ichiron (An essay on the principle of population)*. Dojinsha. https://dl.ndl.go.jp/pid/979690
23. CIPPF: Commission for the Investigation of Problems of Population and Food. 1931. *Jinkō shokuryō mondai chōsakai yōran (Directory of the Commission for the Investigation of problems of population and food)*. Tachi Archive no. 10121607.
24. CIPPF: Commission for the Investigation of Problems of Population and Food. 1928. *Sanji Seigen ni Kansuru Chōsa (A survey on birth control)*. Tachi Archive PDFY10121605.
25. CIPPF: Commission for the Investigation of Problems of Population and Food. 1930. *Jinkō Shokuryō Mondai Chōsakai Jinkō bu tōshin setsumei (Explanation on the report of population section, the Commission for the Investigation of Problems of Population and Food)*. Tachi Archive PDFY091105008.
26. FIRPP: Foundation-Institute for Research of Population Problems. 1937. Institute for the research of population problems in Japan - History, constitutions and activities. https://www.ipss.go.jp/history/foundation/PDF/1937%20HISTORY%20CONSTITUTIONS%20AND%20ACTIVITIES.pdf
27. Nagai, Toru. 1960. *Wagakuni ni okeru jinkō mondai ni kansuru chōsa kenkyū kikan no raireki ni tsuite (History of the organizations for research and study in Japan concerning population problems)*. Annual reports of the Institute of Population Problems, special number in commemoration of the 20th anniversary of the establishment of the Institute, Institute of Population Problems, Ministry of Health and Welfare, Tokyo, Japan.
28. Takaoka, Hiroyuki. 2011. *Sōryokusen Taisei to "Fukushi Kokka" - Senjiki Nihon no "Shakai Kaikaku" Kōsō (Total war regime and "welfare state" - War time Japan's "social reform" plan)*. Iwanami Shoten.
29. Statistical Society of Tokyo. 1937. Nihon jinkō mondai kenkyū iinkai no sōritsu (Creation of the committee on Japan population problem research). *Journal of the Statistical Society of Tokyo* 673:70–72. https://dl.ndl.go.jp/pid/10998684.

30. FIRPP: Foundation Institute for Research on Population Problems. 1938. Dai ikkai jinkō mondai zenkoku kyōgikai hōkokusho (The report of the first National Council on population problems). *Jinkō mondai shiryō (Population Problem Material)* 30. https://dl.ndl.go.jp/pid/1150934
31. IPP: Institute of Population Problems. 1940a. *Sina Jihen ni yoru shusshō oyobi shibō no henka (Change of births and deaths caused by China incident).* Tachi Archive PDFY090212101.
32. IPP: Institute of Population Problems. 1940b. *Jinko seisaku yōkō dai ichiji (The outline for the population policy, the first version).* Tachi Archive PDFY09110501.
33. Statistics Bureau. 1938. *Vital statistics 1937.*
34. Tachi, Minoru. 1946. *Wagakuni genka no jinko mondai (The present population problems of Japan).* Report of Central Unemployment Countermeasures Committee, Tachi Archive PDFB5041P4602.
35. Nakagawa, Tomonaga. 1940. Shōrai jinkō no keisan ni tsuite (On the calculation of future population). *Journal of Population Problems (Jinkō Mondai Kenkyū)* 1 (2): 1–13.
36. Cabinet. 1941. *Jinko Seisaku Kakuritsu Yoko ni Kansuru Ken (Regarding the outline for the establishment of population policy).* JACAR (Japan Center for Asian Historical Records), National Archives of Japan. https://www.jacar.archives.go.jp/das/meta/A03023595500
37. Sanitary Bureau of the Home Department. n.d. The sixth annual report of sanitary bureau. https://dl.ndl.go.jp/pid/836657
38. Hayami, Akira, and Miyoko Kojima. 2004. *Taishō demogurafi-rekishi jinkō gaku de mita hazama no jidai (Taisho demography-the in-between era seen from historical demography).* Bunshun Shinsho 358, Bungeishunjū.
39. Morita, Yuzo. 1954. On the accuracy of demographic statistics of Japan. *Archives of the Population Association of Japan* 3:27–33. https://doi.org/10.24454/apaj.3.0_A27.
40. Tachi, Minoru. 1960. *Formal demography.* Tokyo: Kokon Shoin.
41. Aoki, Hisao. 1967. *Selected statistics concerning fertility regulation in Japan.* Institute of Population Problems Research Series, no. 181, Institute of Population Problems, Ministry of Health and Welfare, Tokyo, Japan. https://www.ipss.go.jp/syoushika/bunken/data/pdf/J08357.pdf
42. Okazaki, Ayanori. 1955. Nihon ni okeru yūsei seisaku to sono kekka ni tsuite (Eugenic policy and its consequences in Japan). *Journal of Population Problems (Jinkō Mondai Kenkyū)* 61:1–7.
43. PAJ: Population Association of Japan. 2002. 50 years history of the population association of Japan.
44. Sugita, Nao. 2009. Historical studies on population policy in Japan: Focusing on Toru Nagai. *Journal of Economics* 110 (1): 54–78. https://dlisv03.media.osaka-cu.ac.jp/il/meta_pub/G0000438repository_KJ00005680179.
45. FIRPP: Foundation Institute for Research on Population Problems. 1946. *Sanji seigen ni kansuru yoron no dōkō (Trend of public opinion on birth control).* Tachi Archive PDFY09111128.
46. IPP: Institute of Population Problems. 1947. *Sanji seigen jittai chōsa kekka no gaihō (Overview of the birth control survey results).* Research Material No. 21. https://www.ipss.go.jp/syoushika/bunken/data/pdf/J08180.pdf
47. Okazaki, Yoichi. 2002. The 50 years of National Survey on family planning (scientific information). *Journal of Population Problems (Jinkō Mondai Kenkyū)* 31:103–111. https://doi.org/10.24454/jps.31.0_103.
48. PAJ: Popoulation Association of Japan. 1952. Archives of the Population Association of Japan. No. 1. https://doi.org/10.24454/apaj.1.0_App1
49. Homei, Aya. 2023. *Science for governing Japan's population.* Cambridge University Press. https://www.cambridge.org/core/books/science-for-governing-japans-population/E7A44D3331E27F04452A5C42CCF2354C.
50. IPP: Institute of Population Problems, Ministry of Health and Welfare. 1955. II Suikei Shorai Jinko (Population projection). *Journal of Population Problems (Jinkō Mondai Kenkyū)* 62:80–90. https://www.ipss.go.jp/syoushika/bunken/data/pdf/14206805.pdf.

51. IPP: Institute of Population Problems, Ministry of Health and Welfare. 1959. *Danjo Nenrei betsu Suikei Jinko (Population projection by sex and age)*. Kenkyu Shiryo No. 118. https://www.ipss.go.jp/syoushika/bunken/data/pdf/J08287.pdf
52. Aoki, Hisao. 1970. A general view of fertility and its regulation in Japan. *Journal of Population Problems (Jinkō Mondai Kenkyū)* 114:5–20. https://www.ipss.go.jp/syoushika/bunken/data/pdf/14212002.pdf.
53. Okazaki, Yoichi. 1970. An analysis of socio-economic factors affecting fertility in Japan. *Journal of Population Problems (Jinkō Mondai Kenkyū)* 114:21–34. https://www.ipss.go.jp/syoushika/bunken/data/pdf/14212003.pdf.
54. Tachi, Minoru. 1969. *Population problems in Japan*. Institute of Population Problems Research Series, No. 190. https://www.ipss.go.jp/syoushika/bunken/data/pdf/101786_1.pdf
55. ACPP: Advisory Council of Population Problems. 1969. Jinko Mondai Shingikai Chukan Shinto (Intermediate report of the advisory council of population problems). *Journal of Population Problems (Jinkō Mondai Kenkyū)* 112:67–70. https://www.ipss.go.jp/syoushika/bunken/data/pdf/14211808.pdf.
56. Superstition Research Council, Ministry of Education. 1949. *The present state of superstition*. Tokyo: Gihodo. https://dl.ndl.go.jp/pid/2982577/1/52.
57. Kurosu, Satomi. 1992. Hinoeuma in Koka 3 (1846): Regional variations of culture and population. *Bulletin of International Research Center for Japanese Studies* 6. https://doi.org/10.15055/00000900.
58. Kure, Ayatoshi. 1911. Sengo no Shusshou Fu Hinoeuma no Meishin (Post-war fertility, with the superstition on fire-horse year). *Journal of the Statistical Society of Tokyo* 363:356–360.
59. Ministry of Health and Welfare. 1969. [Tokushū] Shōwa 41 nen no shusshō genshō ni tsuite ([special report] on the fertility decline of Shōwa 41 (1965)). *Vital Statistics 1966 Japan* 1:68–77. https://dl.ndl.go.jp/pid/3048831.
60. Kodansha. 1965. Hinoe-uma no akachan wo undara taihen (It would be hard to give birth to a baby girl in the fire-horse year). *Young Lady* 1965:128–135.
61. Sakai Hiromichi. 1995. The study of socio-demographic behavior relevant to 'Hinoe-Uma' in 1966. *Journal of Population Problems (Jinkō Mondai Kenkyū)* 18:29–38.
62. ACPP: Advisory Council of Population Problems. 1974. *Nihon jinkō no dōkō - Seishi jinkō wo mezashite (Trend of Japanese population: Towards a stable population)*. Ministry of Finance Printing Bureau. https://dl.ndl.go.jp/pid/11944162.
63. United Nations, Department of Economic and Social Affairs, Population Division. 2024. *World population prospects 2024*. Online Edition. https://population.un.org/wpp/
64. Ono, Taichi. 2014. *Shakai hoshō, sono seisaku katei to rinen (Social security, its policy process and philosophy)*. Shakaihoken Kenkyusho.
65. IPSS: National Institute of Population and Social Security Research. 2012. Report on the fourteenth Japanese National Fertility Survey in 2010, Volume I. Survey Series No. 29. https://www.ipss.go.jp/syoushika/bunken/DATA/pdf/207616.pdf
66. Cabinet Office. 2004. Heisei 16 nen ban Shōshika shakai hakusho (Annual report on the declining birthrate 2004). https://warp.da.ndl.go.jp/info:ndljp/pid/12772297/www8.cao.go.jp/shoushi/shoushika/whitepaper/measures/w-2004/html_h/index.html
67. House of Councillors. 2015. 189th House of Councillors Budget Committee Verbatim No. 3. https://kokkai.ndl.go.jp/txt/118915261X00320150203

Chapter 2
Mortality: Epidemiological Transition and Health System Development

2.1 Introduction

Japan boasts one of the world's longest life expectancies, which continued to increase until 2020, the first year of COVID-19. However, due to an ageing population, deaths have become more frequent. The crude death rate surpassed 1% with the number of deaths reaching 1.58 million in 2023, a level comparable to the pre-WWII period (Fig. 2.1).

Since the inception of modern statistics in 1872, the crude death rate fluctuated but remained consistent until the 1910s. The Spanish influenza severely impacted mortality in 1918, after which it decreased. A rapid increase in life expectancy at birth can be observed after WWII until the mid-1950s. Subsequently, from the 1960s, the increase slowed but persisted. This chapter outlines these distinct phases of mortality change, alongside the evolution of mortality statistics and policies.

Mortality policy as a component of population policy, influences both population size and individual lives [1]. However, it is often viewed as a health policy rather than a population policy. A unique characteristic of mortality policy, compared to fertility or migration policies, is its unidirectional nature. Unlike the bidirectional policy options for fertility or migration, decreasing mortality is consistently considered beneficial. Therefore, the central question is how people measured mortality and responded to the problems they perceived.

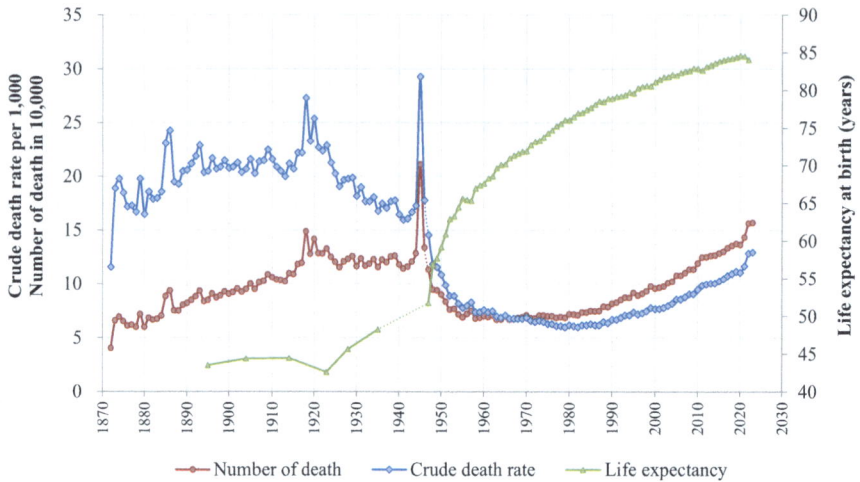

Fig. 2.1 Trend of mortality in Japan. Note: The low 1872 data is due to the limited months counted in that year. Sources: Family register statistics (Home Department) and Annual Report (Sanitary Bureau), Vital Statistics and Life Tables (Statistics Bureau and Ministry of Health (Labour) and Welfare), and Japanese Mortality Database (JMD) (IPSS)

2.2 The Creation of Health Administration and the Cause of Death Statistics in Early Meiji

As in all societies, medicine has existed in Japan for as long as Japanese people have inhabited the country. Early literary mentions of medicine men include the gods Ōkuninushi and Sukunabikona, credited with introducing disease-curing methods to Japan [2]. Subsequently, with the in-flux of people, Korean and then Chinese medicine entered Japan. These influences led to the compilation of medical canons such as Daido Ruijuho in 808 and Ishinpo in 984. Western medicine first arrived with Jesuits and Christianity in the sixteenth century, and reappeared in the mid-eighteenth century through Dutch people, who were the only Westerners permitted contact with Japan via Dejima island in Nagasaki. All these transmissions and accumulated knowledge shaped medical practices where techniques were passed down from father to son and remained within family lines. This knowledge was not publicly disseminated, and there was no philosophy to systematically expand medical practice into a national health system. Nevertheless, based on these existing medical practices, approximately 23,284 doctors were present across the country at the start of the Meiji period [3]. Considering the total population of around 30 million at that time, this meant close to 1 doctor per 1000 population, a level recommended by the WHO.

The Meiji restoration and the subsequent "modernization", or "Westernization", abruptly established a nationally extended health system through the promulgation of Isei, the Medical Ordinance, in 1874. Dr. Nagayo Sensai, as a member of the Iwakura Mission, visited the United States and Europe from 1871 to 1873 and "discovered" as the concept of "health administration" [4]. Upon his return to Japan, he became the head of the Sanitary Bureau, drafted the Medical Ordinance, and implemented public health administration.

Under the Medical Ordinance, the health system was structured to include health personnel education, hospital management, drug administration, and also cause of death registration and infectious disease surveillance. Along with the family register implemented in 1872, doctors nationwide were ordered to examine deaths and report the cause of death. The first cause of death statistics for 1875 were published in the first report of the Sanitary Bureau [3].

The 1875 cause of death statistics covered only Tokyo, Osaka, and Kyoto prefectures for six months, from July to December 1875. The entire country was covered in the subsequent period, from January to June of 1876. During that period, the total number of deaths with a reported cause was 95,689, representing 31.2% of total registered deaths. This proportion increased annually, achieving full coverage within seven years, by 1881. The high density of medical doctors throughout the country can account for this rapid progress. Furthermore, coverage accelerated after 1879, when Cholera struck the archipelago, causing 105,786 deaths. The epidemics likely contributed to achieving full coverage by raising awareness among doctors and the general population about the importance of death registration.

The classification of causes of death appearing in the annual report was nearly identical to Farr's proposed classification discussed at the International Congresses of Statistics [5, 6]. Dr. Nagayo must have been informed about cause of death statistics and their classification during his stay in Europe. Additionally, at the beginning of death registration, a detailed list of disease names used in both Western and Chinese medicine was distributed to doctors nationwide via regional health officers to facilitate cause of death determination [7]. From 1875 to 1901, the classification was slightly modified twice, in 1878 and 1884, remained consistent until 1901. In earlier years, diseases of the digestive organs were the most frequent cause, accounting for around 25% of all deaths, but they were surpassed by diseases of the respiratory organs, including "pulmonary disease" since 1898 (Fig. 2.2). This marked the emergence of the "national disease", tuberculosis. In addition to these, diseases of the nervous system remained one of the main causes of death.

The Statistics Bureau began publishing cause of death statistics in 1899 as a section within vital statistics. The Sanitary Bureau continued its own cause of death statistics using its own classification until 1902, when it was integrated into vital statistics. The Sanitary Bureau's report for that year listed the correspondence between the two different classifications [8]. The Sanitary Bureau's classification comprised 12 categories, presumably based on Farr's classification, while the Statistics Bureau's classification consisted of 46 categories, based on Bertillon's classification or the first International Classification of Diseases (ICD) [9, 10]. Consequently, the Statistics Bureau's vital statistics classification was more detailed and aligned with international standards. However, when the new classification was applied to death statistics, as many as 15% of deaths were categorized as "other diseases" (Fig. 2.3). The new international classification was not entirely compatible with doctors' descriptions of causes of death. The proportion of these unclassifiable deaths remained considerably high for subsequent years.

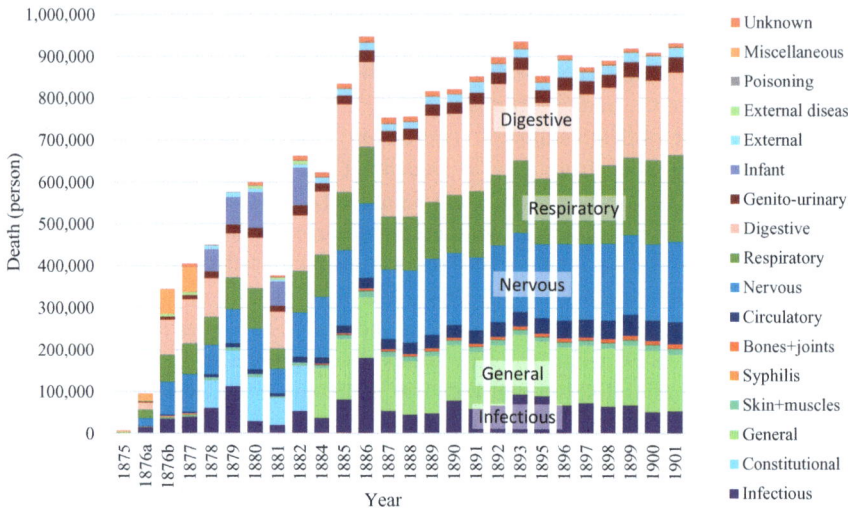

Fig. 2.2 Number of deaths by cause (1875–1901). Note: Year refers to the administrative year from July to June of the following year from 1876b to 1880. 1875 is from July to December 1875, 1876a is from January to June 1876, and 1881 is from July to December 1881. "Constitutional" refers to "whole body" in Japanese likely corresponding to the term used by Farr [5]. Source: Annual Report (Sanitary Bureau)

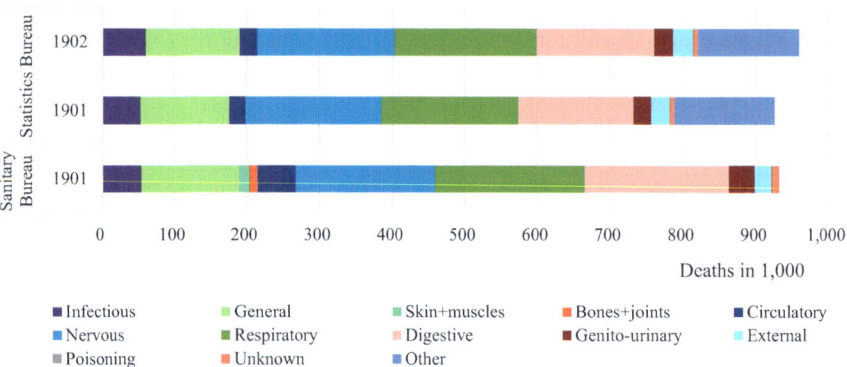

Fig. 2.3 Comparison of causes of death statistics between the Sanitary Bureau and Statistics Bureau. Sources: Annual Report (Sanitary Bureau), Vital statistics (Statistics Bureau)

2.3 The Fights Against Infectious Diseases

2.3.1 *Cholera*

Numerous infectious diseases afflicted Japan [2]. Pre-modern Japan avoided the global bubonic plague pandemic in the fourteenth century, with smallpox being the most severe infectious disease [11]. However, from the nineteenth century, even under the *Sakoku* regime, which literally means "locked up" and limited international contact, Japan became involved in the wave of globalization with the global cholera pandemic, which reached Japan in 1822.

Regarding this 1822 cholera outbreak in Japan, two theories exist concerning the transmission route. The more probable route was from the Chosun Dynasty (Korea) to Tsushima, the island closest to Korea, and then to Yamaguchi; the other was from Batavia to Nagasaki, and then from Kyushu to eastern Japan [12]. During the Edo period, Nagasaki Dejima was considered the sole entry point to Japan, yet international relations were maintained through Tsushima and Ryukyu, with an average of 25.9 vessels per year from 1635 to 1852 [13]. Cholera may have entered Japan in 1822 via the first route, through Tsushima. In any case, this initial cholera outbreak did not reach Edo and disappeared within the year. Earlier in that year, doctors in Edo were informed by Dutch doctors about the cholera outbreak in Batavia. This information was transmitted through letters among doctors in different parts of Japan but effective countermeasures were difficult to implement. Doctors believed it was caused by miasma, and people resorted to prayers and amulets [14]. The magnitude of the pandemic could only be gauged from fragmented descriptions in personal letters. The death toll rose to 3000 in Yamaguchi and several thousand in Osaka.

The second cholera pandemic arrived in Japan in 1858. This time, the entry point was clearly established: the American USS Mississippi frigate entered Nagasaki on July 1, 1858, and a crew member suffering from cholera disembarked. Transmission was rapid; 30 cholera infection cases occurred in Nagasaki on the same day, then spread eastward, reaching Edo in August [12]. The country was on the verge of transition, and tightly regulated internal migration might have loosened at that time. Without a proper national governmental body in charge of public health, some regional governments (*Han*) informed people about precautions and counted the number of cholera cases and deaths. A specialized hospital was established in Nagasaki. People were advised to clean their homes, eat and drink moderately, and avoid blue-skinned fish, cucumber, or watermelon among other foods. In case of contracting the disease, a special medicine made of cinnamon and ginger was recommended, as well as drinking diluted alcohol or applying it to the skin. There were 221 deaths in Nagasaki, 10,000 in Osaka, and 30,000–40,000 deaths in Edo [12].

Entering the Meiji Era, several waves attacked Japan, the first in 1877, followed by major outbreaks in 1879 and 1886 (Fig. 2.4), which were instrumental in the development of public health [4, 16]. Cholera prevention notifications and regulations were issued and revised multiple times. Doctors nationwide were instructed to report cholera cases to local authorities in 1875. As Fig. 3.4 demonstrates, the epidemiological surveillance system was quickly established, and the number of cases and deaths was published in 1876 by the Sanitary Bureau.

During this period, proper quarantine was not possible due to unequal treaties between Japan and Western countries. One of the earliest cholera epidemics in 1877 involved an infected crew of a Great Britain warship anchored in Nagasaki. Representatives of countries with extraterritorial rights, notably Great Britain, Germany, and France, rejected the proposed sea quarantine regulations. The effective quarantine of international traffic had to wait until 1899 when treaty revision was achieved.

However, the unequal treaty was not the sole cause of the cholera pandemic's spread. Internal migration of Japanese, especially military soldiers, hastened its dissemination. The 1877 epidemics were carried by soldiers who fought in the *Seinan* War and returned to Kansai and Tokyo. Similarly, the 1895 epidemic was caused by the Sino-Japanese War through infected returning soldiers from China to Japan.

The low level of hygiene was a significant contributing factor to cholera propagation. Until the first modern water supply system was completed in Yokohama city in 1887, the entire country had only a few water supply systems dating from the Edo period, and a modern sewage system was almost non-existent. The repetitive spread of cholera drew public attention, and national policies on modern water supply systems commenced. In 1887, the Central Hygiene Association submitted a proposal for the construction of a modern water supply system in Tokyo, and a Cabinet decision on the water supply system was issued. This decision granted a national subsidy to cover one-third of the construction cost for three metropolitan prefectures and five major ports. Three years later, the first law on water supply was enacted. The City Institution Law promulgated in 1888 facilitated the installation of the water supply systems in cities. Extending the water supply system throughout the country, especially in urban areas, took time, but it reduced acute infectious diseases (Fig. 2.4).

2.3.2 Tuberculosis

The official statistics began using the term "tuberculosis" in 1899, when there were 67,346 deaths (Fig. 2.4). Before that, tuberculosis and similar diseases were referred to as "pulmonary disease (*Hai-byo*)", and it became a single entry in the cause of death statistics in 1883, a year after Dr. Koch identified Mycobacterium tuberculosis and proved it to be an infectious disease [17, 18]. The pulmonary disease death toll was only 3 in 1885 but surged to 36,138 the following year. This increase must have been due to increased recognition by doctors or statisticians.

Unlike cholera, Japanese people had suffered from tuberculosis since antiquity; the oldest evidence of tuberculosis was found on bones excavated from a Yayoi period tomb [19]. As no such evidence found in bones from earlier Jomon period, it

2.3 The Fights Against Infectious Diseases

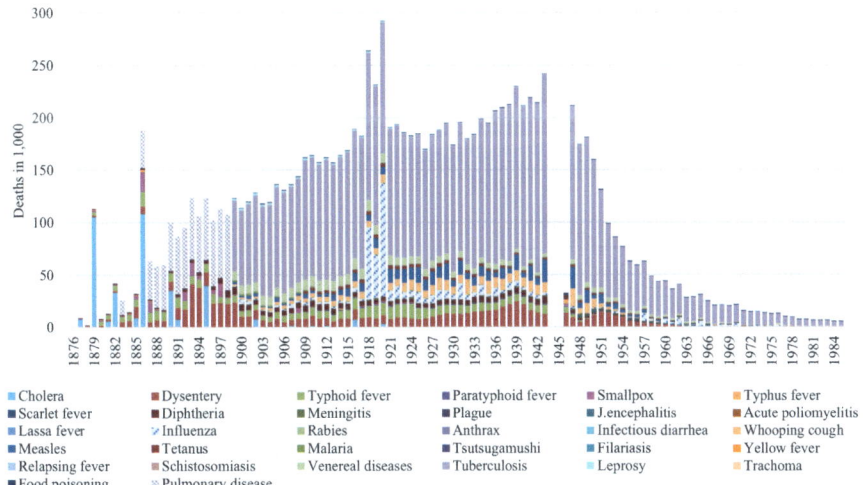

Fig. 2.4 Deaths by Infectious Diseases (1876–1985). Source: Annual Report (Sanitary Bureau), Vital Statistics (Statistics Bureau and Ministry of Health (Labour) and Welfare), complied in Japan Statistical Association [15]

is assumed that tuberculosis entered Japan with the migration flow from China and Korea. The oldest Japanese medical textbook, "*Ishinpō*", written by Yasuyori Tanba in the Heian period in the tenth century, dedicated its thirteenth chapter to tuberculosis [20]. The Heian period mummified remains of Lord Hidenari Fujiwara, who died in 1187, also exhibit spinal caries, a sign of tuberculosis infection [21]. During the Edo period, the disease appears to have been more prevalent. The great fire of Meireki in 1657, which killed 100,000 residents of Edo city, was allegedly attributed to the burning dress worn by three women who died of tuberculosis. The quantitative description of tuberculosis first became available in the mid-nineteenth century when Dr. Gendo Honma described 55 tuberculosis patients over a year and a half from the spring of 1847 [22]. Retrospective historical demographic research showed that tuberculosis was the tenth cause of death, comprising 1.5% of total deaths with a death rate of 62.7 per 100,000 population for the period from 1831 to 1840 in a village of the Hida region [21, 23].

Entering the Meiji era, tuberculosis was identified as a public health concern. An epidemiological survey was conducted in urban areas in 1882, and the number of deaths caused by "pulmonary death", a slightly larger definition than tuberculosis, counted 2355 in Tokyo on that year, corresponding to 7% of total deaths [24, 25]. The following year in 1883, the number of deaths caused by "pulmonary deaths" began to be published in the annual report of the Sanitary Bureau. A tuberculosis treatment laboratory was established at the University of Tokyo, and the first sanatorium for tuberculosis patients opened in Suma, Hyogo, in 1889. However, the death toll continued to increase linearly (Fig. 2.5). During this period, higher priority was given to acute infectious diseases such as cholera, dysentery, or smallpox,

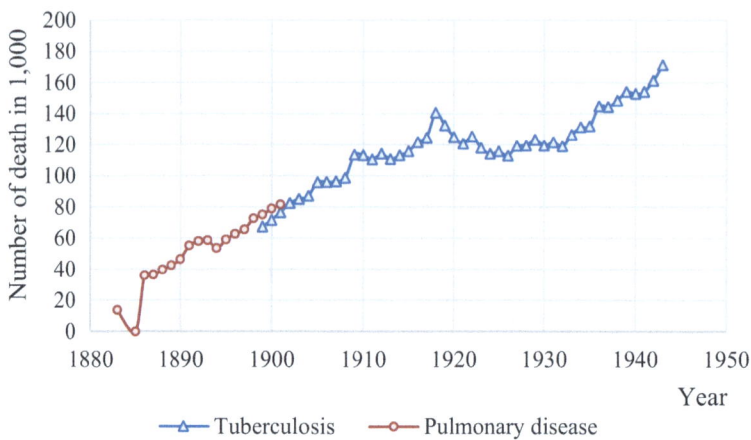

Fig. 2.5 Number of deaths by tuberculosis. Sources: Pulmonary disease in Annual Report (Sanitary Bureau), Tuberculosis in Vital Statistics (Statistics Bureau)

and "pulmonary disease" was not included in the infectious disease surveillance system; consequently, there were no statistics on the number of cases.

Around the turn of the century, from the 19th to the 20th, when cholera began to be a controllable disease, the fight against tuberculosis gained momentum. The Act on Prevention of Infectious Diseases and Disinfection Method at School was stipulated in 1898, followed by the Act on Cattle Tuberculosis Prevention in 1901, and finally, the Ministerial Order on the Prevention of Pulmonary Tuberculosis in 1904. Dr. Robert Koch visited to Japan in 1908, gave lectures throughout Japan for two months, had an audience with the emperor, and raised the awareness of the Japanese people with wide media coverage. The Japanese Anti-Tuberculosis Association was established in 1913, and the Act on Tuberculosis Prevention was passed in 1919.

Japanese tuberculosis mortality patterns differed from those in other countries, being higher for young women than for young men. This pattern accelerated around 1897. The massive internal migration of young women to work in spinning factories was a source of the rapid spread of tuberculosis throughout the country [26]. Recognizing a problem is halfway to solving it. The Act on Factory, enacted in 1916, protected the factory workers, especially minors. The health checkups stipulated in the law helped prevent the spread of tuberculosis.

Despite these interventions, the tuberculosis death toll continued to rise until it peaked at 140,747 deaths in 1918, after which it began to decline (Fig. 2.5). How can this distinct trend be explained? Firstly, the increasing trend from the beginning of the statistics could be due to the growing recognition of tuberculosis. During the same period, the number of deaths with unknown causes declined. The second factor was the Spanish Influenza. The pandemic arrived in Japan in August 1918 and caused the first sharp death peak in November. Tuberculosis deaths also peaked in November, as many tuberculosis patients were more vulnerable to the Spanish Influenza [27]. The decrease in tuberculosis from 1919 could be attributed to the "harvest effect" of Spanish influenza, which killed vulnerable tuberculosis patients earlier than expected [24], or to general behavioral changes to prevent infection.

Since the 1930s, tuberculosis mortality increased again. This increase is attributed to the war, with many young men drafted into the military, which was also a field for communicable diseases. The number of tuberculosis deaths among men surpassed that of women in 1931. Drastic eradication had to wait until after WWII.

2.3.3 Epidemic Cold (Spanish Influenza)

The spread of Spanish influenza, known as Epidemic Cold (Ryūkōsei Kanbō) at the time, began in Japan in August 1918. According to a special report by the Sanitary Bureau, it occurred in three waves, peaking in November 1918, January 1920, and March 1921, resulting in a total of 23,804,673 cases and 388,727 deaths [28]. According to vital statistics for the corresponding period from August 1918 to July 1921, the number of deaths caused by the Epidemic Cold was 225,714, significantly lower than the Sanitary Bureau's report. This discrepancy was due to confusion regarding cause of death certification. Many potential Epidemic Cold deaths were attributed to pneumonia, other respiratory diseases, and unknown causes, especially during the first wave. Additionally, the Spanish Influenza affected vulnerable individuals who were already ill, leading to a sudden increase in deaths caused by tuberculosis, as mentioned earlier, but also by cerebrovascular diseases, heart diseases, eclampsia, convulsions or senility. Hayami [29] estimated 453,152 deaths caused by the Epidemic Cold, based on excess mortality from related causes.

Unlike the COVID-19 pandemic a century later, the Epidemic Cold affected Japan at a similar level to Western countries. The monthly number of deaths surged in November 1918 and January 1920. Approximately 80% of this increase was due to the Epidemic Cold and pneumonia (Fig. 2.6).

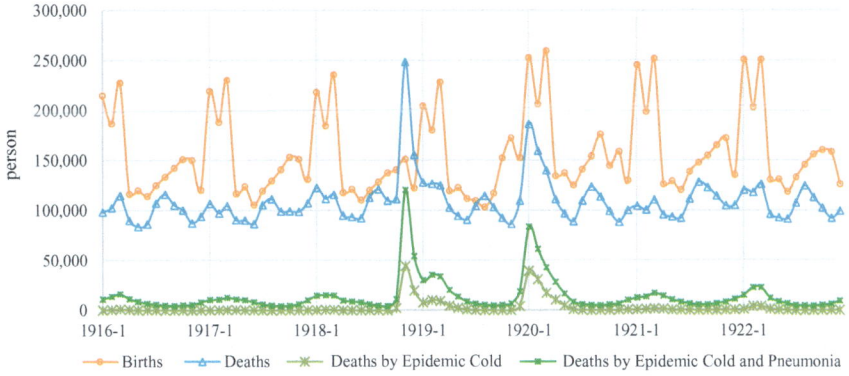

Fig. 2.6 Monthly number of births and deaths from 1916 to 1922. Source: Vital Statistics (Statistics Bureau)

At the onset of the Epidemic Cold pandemic, the Sanitary Bureau was preoccupied with other infectious diseases. Infectious Meningitis, with a high case fatality rate of around 60%, was designated as an infectious disease in 1918 [30]. The Tuberculosis Prevention Act was enacted in 1919. In contrast to meningitis or tuberculosis, the case fatality rate of the Epidemic Cold was low, but it caused a tremendous number of deaths due to the extremely high number of cases [31]. Unlike the previous influenza pandemic in 1890, the central government implemented various measures for prevention and treatment. These measures included advocating mask-wearing, gargling, closing schools, promoting vaccination, and arranging hospital facilities. The governmental budget for health increased fourfold, from 570,189 yen in 1917 to 2,196,461 yen in 1920.

Fertility was also affected by the pandemic. Nine months after the first mortality peak in November 1918, the number of births dropped in August 1919. However, in subsequent months, the number of births increased and even surpassed previous years' figures (Fig. 2.6). As a result, the annual number of births surged in 1920, which later contributed to the sudden increase of centenarians a century later in 2020.

2.4 1920: The Onset of Mortality Decline

2.4.1 *The Quality of Statistics for Infant Mortality*

Mortality began to decline in 1920 (Fig. 2.1). There were debates about whether this represented a genuine trend or was due to improvements in registration. Similar to fertility data, as discussed in the previous chapter, the number of deaths was also assumed to be under-registered, and life expectancies from 1895 to 1914 were considered too high by later demographers [32–35]. According to them, it was impossible for Japan's life expectancy to be higher than that of more advanced Western countries before 1920.

However, the mortality statistics system did improve from the late nineteenth century. The Graveyards and Burial Control Regulation, stipulated in 1884, urged people to register deaths; otherwise, they could not properly conduct funerals for the deceased. In 1886, the Family Register Act was amended, and from 1890 onward, cause-of-death data by single age became available in the Sanitary Bureau's annual report. The Midwife Regulation stipulated in 1899 contributed to the registration of stillbirths. Births and deaths were counted by both the Directorate of Family Register and the Sanitary Bureau within the Home Department until 1898, when the Statistics Bureau assumed responsibility for compiling vital statistics. This was a groundbreaking change, as the tabulation format report was replaced by a single information sheet for each life event. The Statistics Bureau had the opportunity to individually control the quality of

2.4 1920: The Onset of Mortality Decline

original birth and death registrations and create detailed tabulations according to their needs. Concurrently, the Statistics Bureau produced "static statistics", or registered-based censuses in modern terms, every five years to obtain detailed population figures by age and sex until they were replaced by a proper population census in 1920.

Infant mortality increased rapidly until 1890, but this trend did not continue thereafter, then it turned to a continuous decline from 1920 (Fig. 2.7). The Sanitary Bureau considered infant mortality before 1890 an "exception" [36]. Yuzo Morita, who became the head of the Statistics Bureau in 1947, believed that data from 1890 was already reliable [37]. Takase [38] found that the 1890 registered population could be accurately reconstructed from the 1920 census population by backward annual addition and subtraction of births and deaths between those years. If we accept these observations by contemporary health administrators and later scholars, the rise in infant mortality until 1890 was due to the registration improvement, and after 1890, the mortality data reflects the real trend as it became trustworthy. Then, we need to know what made the infant mortality change from 1890, especially the decline from 1920.

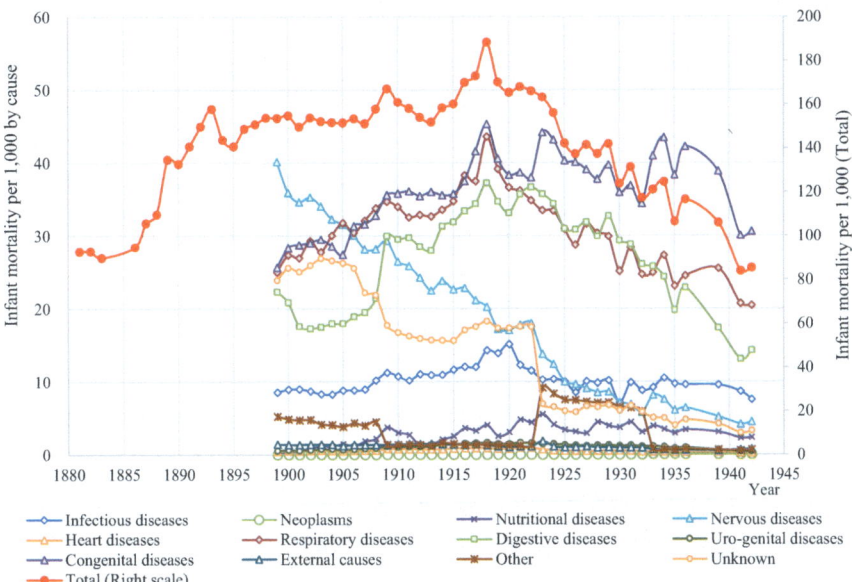

Fig. 2.7 Infant mortality rate by cause. Note: Sudden changes are caused by classification changes in 1909, 1923, and 1933. Tuberculosis is included in infectious diseases. Sources: Annual Report (Sanitary Bureau), Vital Statistics (Statistics Bureau)

2.4.2 Infant Mortality by Cause of Death and by Prefecture

Using vital statistics, we can observe infant mortality by cause and by 47 prefectures from 1899. By cause, the sharp rise in infant mortality in 1918 was related to the "Epidemic Cold (Spanish Influenza)", with a substantial increase in respiratory and infectious diseases. However, at the same time, other diseases, notably congenital and digestive diseases, contributed to the increase in infant deaths. Subsequently, mortality began to decline steadily (Fig. 2.7), primarily due to reductions in both respiratory and digestive diseases, the two main causes of infant deaths, while deaths caused by congenital diseases remained high. Throughout the period from 1899 to 1942, there was a consistent decline in nervous diseases and unknown causes.

Geographically, the declining trend from 1920 is observed across all 47 prefectures, with a converging trend from 1920 to 1943. However, a more radical decrease is seen in metropolitan prefectures: Osaka, Kyoto, and Tokyo (Fig. 2.8). Okinawa, the southernmost island prefecture with a distinct culture from mainland Japan, shows a distinctly lower level of mortality, which could be due to under-registration.

2.4.3 The Determinants of Infant Mortality Decline

Assuming the decline in infant mortality was a substantive change rather than a mere statistical artifact, what were the underlying causes that led to improved child survival around 1920? The determinants of this decline have been extensively discussed, and we can now examine these arguments with data.

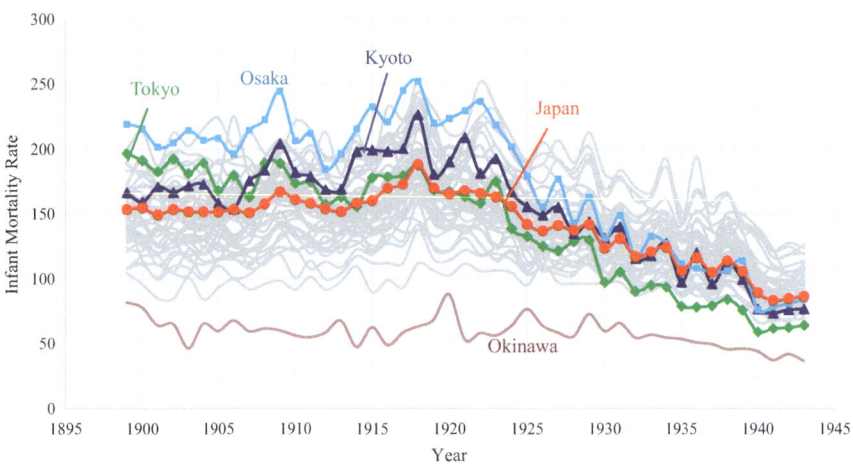

Fig. 2.8 Infant mortality rate by prefecture (1899–1943). Source: Vital Statistics (Statistics Bureau)

2.4 1920: The Onset of Mortality Decline

Regarding the role of health system, some scholars, such as Nishida [39], contend that medical techniques were not a contributing factor, given that the decline preceded the advent of effective interventions like tuberculosis chemotherapy. However, even in the absence of advanced medical technology, the health system underwent significant development during this period. The Health and Hygiene Investigation Commission, established in 1916, designated infant and child mortality as its primary investigative theme, before the topics such as tuberculosis, sexually transmitted diseases, leprosy, mental diseases, clothing-food-housing, rural hygiene, and statistics [40]. The Commission's activities continued until 1939, consistently advocating for health system improvements [41–44].

A parallel policy improvement occurred in social welfare. With the establishment of the Social Bureau in 1920 within the Home Department, the traditional "charity work" was replaced by the government-supported "social work". Higami [45] suggests that community-based social work had a discernible impact on reducing infant mortality. While measuring the precise effectiveness of these activities is challenging, government budget data reveals a sharp increase towards 1920 for health, 1921 for social work (Fig. 2.9a). The sharp rise in health spending in 1920 may have been a direct response to the Spanish Influenza, but thereafter, the budget for health and social work remained elevated. The confluence of the Spanish Influenza pandemic and the concurrent development of health and social work policies could plausibly account for the decline in infant mortality.

The role of midwives was also an essential factor in preventing infant deaths. Well-trained midwives provide proper prenatal and postnatal care, as well as education for mothers, which would prevent infant mortality. Traditional birth attendants in the Edo period, who also practised abortion and/or infanticide, were banned and replaced immediately after the Meiji restoration. The Midwife Regulation (Sanba Kisoku), stipulated in 1899, paved the way for promoting the status of midwives. However, the number of midwives increased steadily since 1899. Hence, this factor does not explain the turnaround in mortality in 1920 (Fig. 2.9b).

Other factors contributing to the decline in infant mortality are at the individual level. If the education level is higher, especially for women, better-informed care for babies should lower child mortality. From the beginning of the Meiji era, education was another sector to which the government attached great importance, and women's basic education expanded rapidly. The school enrollment of girls aged 6 to 12 increased rapidly from 15.1% in 1873 to almost universal at 97.4% in 1910 (Fig. 2.9c). Considering the average age of marriage for women was 24.2 years old in 1920, it is plausible that fully educated women at that point could have contributed to the lowering of infant mortality. However, as the school enrollment trend was gradual, it does not clearly explain the sudden change in infant mortality in 1920.

Besides health system and individual factors, environmental factors should be considered. For example, the quality of drinking water is an important determinant in mitigating digestive diseases, including diarrhea. Takemura [46] suggested that piped water disinfection, introduced in 1921 using unused chlorine originally intended for the Manchuria invasion, triggered the infant mortality decline. This

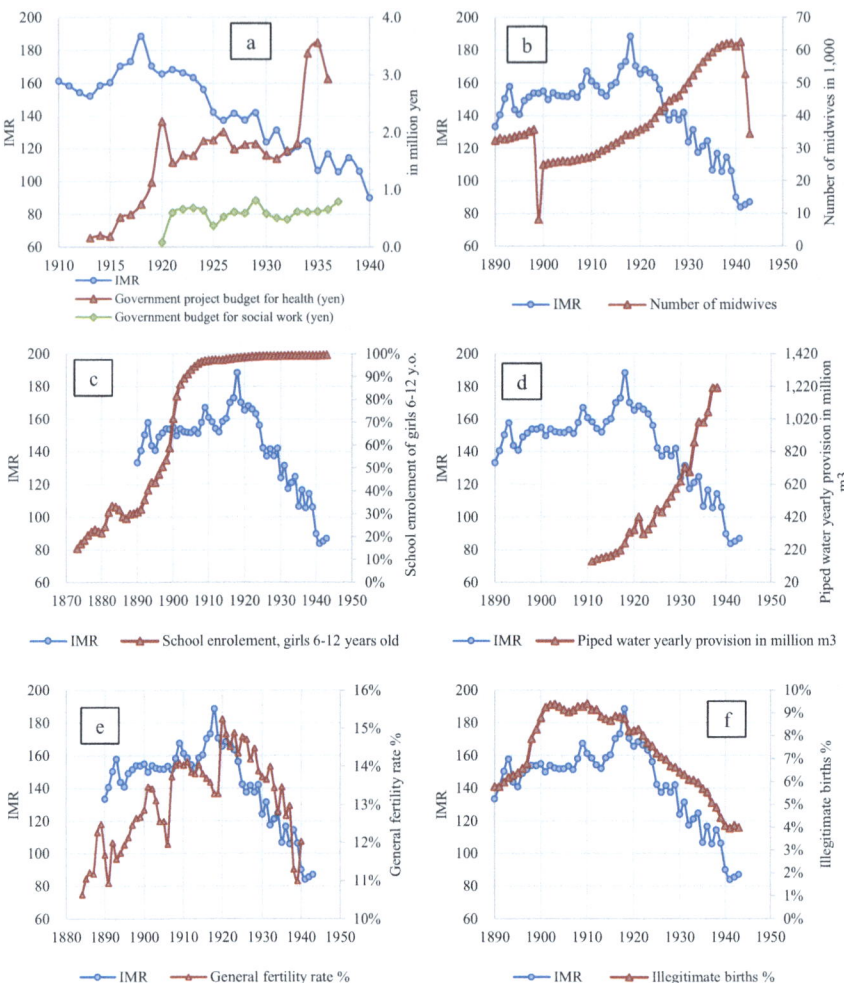

Fig. 2.9 The Infant mortality rate (IMR) and related indicators. Note: Government budget for social work is the sum of both ordinary budget (including salary of civil servants) and project budget. Government project budget for health is not including ordinary budget due to the data unavailability. Sources: Vital Statistics (Statistics Bureau) for infant mortality rate, general fertility rate and illegitimate birth, Statistical Yearbook (Statistics Bureau) for government budget and piped water, Historical Statistics of Japan (Japan Statistical Association) for number of midwives and school enrollment

process was facilitated by Shinpei Goto, a medical doctor specializing in microbiology who studied under Dr. Koch in the German Empire, who was the Foreign minister at the time of the Manchuria invasion, and subsequently the Mayor of Tokyo from 1920. The growth of piped water provision began much earlier (Fig. 2.9d), the statistics on the coverage of chlorine disinfection during those days are unavailable, and the commonly held view is that chlorination of piped water gained full momentum after WWII [47]. However, the rapid decline of infant mortality after 1920 was notable in Tokyo, when Dr. Goto was the Mayor; this factor cannot be dismissed as a reason for the infant mortality decline.

The final factor to consider is fertility itself. The relationship between fertility and infant mortality is bidirectional. If a child survives, there is no need to have additional babies, thus fertility decreases. On the other hand, as Dr. Hiroshi Maruyama, a dedicated epidemiologist who fought against infant mortality lamented how many infant deaths were caused by parental indifference [48], high fertility with unwanted pregnancies could lead to child deaths and thus high infant mortality. The decline in infant mortality from 1920 is well linked with the general fertility rate (the number of births divided by the female population aged 15–49) (Fig. 2.9e). The relationship is even more apparent when observing the legitimacy of birth. In Osaka, one-third of illegitimate births resulted in death [49], and nationwide, the infant death rate for illegitimate children was 218 per 1000 births, significantly higher than that of legitimate children, 155, according to Vital Statistics in 1910. The proportion of illegitimate births among total births mirrored the trend of the infant mortality rate quite closely (Fig. 2.9f). The trend of the illegitimate birth rate has been attributed to the change in the old practice where couples registered marriage only after having their first child [50]. However, this factor alone does not fully explain why it increased until 1900, and it conincided with the trend of infant mortality. Both the illegitimate birth rate and infant mortality rate rose around 1910, declined thereafter, increased or stagnated until 1918, and then began to decline continuously.

The increase and decrease in the illegitimate birth rate could be due to the spread of the notion of marriage registration and the importance of having a child within wedlock. Additionally, the decrease in illegitimate births starting in 1910 could be due to increased awareness of birth control. Neo-Malthusian theory entered Japan following the International Neo-Malthusian Conference held in Paris in 1900. Sadao Oguri published a book on social improvement through birth control in 1903. Mrs. Margaret Sanger visited Japan in 1922, and the Working Group on Birth Control was established in Tokyo in the same year. Already, contraceptives such as condoms, pessaries, washing devices, and metal uterine plugs were available and commercialized at the time [51]. Although no nationwide survey on contraceptive prevalence existed then, the first figure was 28.6% in 1950 [52]. It is plausible that birth control methods began to spread gradually around 1920. This could be a powerful explanation for the decline in the illegitimate birth rate and the reduction of unwanted children, which led to the decline in the infant mortality rate.

2.5 The Path to Universality: Health Insurance Expansion

Universal Health Coverage (UHC) was achieved in Japan in 1961. However, its foundations were laid during World War II. The promulgation of health insurance laws, starting with the Health Insurance Law in 1922, followed by the National

Health Insurance Act in 1938 and related revisions, expanded both the types of medical insurance and the number of subscribers. By 1944, the public health insurance system covered 70% of the population (Fig. 2.10). Coverage increases to 96% when including dependents who received benefits.

The process of UHC expansion can be divided into three phases. The first phase was during the 1920s, when the rise of communism and socialism, inspired by the Russian Revolution, spurred the formation of labour unions and pressured the government to protect workers. This phase is characterized by the establishment of the Social Bureau of the Home Department in 1920 and the promulgation of the Health Insurance Law in 1922. The second phase occurred after the outbreak of the Sino-Japanese War in 1937, followed by the National Health Insurance Law in 1938, the Cabinet Decision on the "Outline for Establishing Population Policy" in 1941, and then the National Health Insurance Law in 1942. The "Outline for Establishing Population Policy" was renowned for its fertility-related ordinances, but it also included chapters concerning population health improvement. It explicitly aimed to expand the health insurance system and promote maternal and child health. This period saw the inauguration and expansion of what would later be called "District Health Systems" (WHO 1995), aiming to cover all villages with a health center and health personnel. The third phase was after WWII, when UHC was achieved in 1961, following the devastation of pre-war health insurance institutions due to post-war hyperinflation.

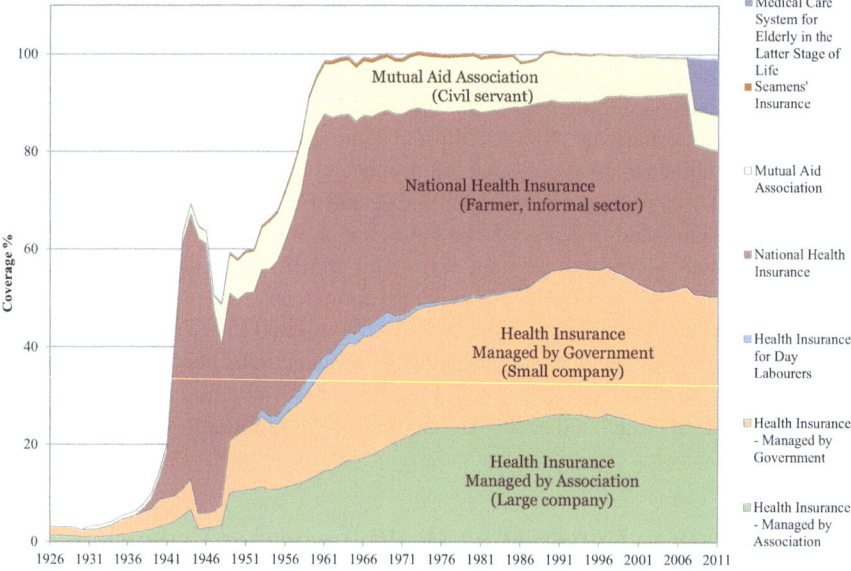

Fig. 2.10 Health Insurance Coverage in Japan (1926–2011). Note: Before 1949, the number of dependents of the insured was not included in the statistics. Hence, the total number of beneficiaries is substantially larger from 1949, when the payment started to cover the dependents. Sources: Ministry of Health and Welfare [53], Japan Statistical Association [15], IPSS [54], Takagi [55]

2.5 The Path to Universality: Health Insurance Expansion

Laws on health insurance systems were developed by national government bodies and evolved through the period, starting from the Ministry of Agriculture and Commerce, then passing to the Social Bureau of the Home Department, and finally, the Ministry of Health and Welfare and the Insurance Agency created in 1938. However, municipal governments and health centers, as well as community organizations such as industrial associations, women's associations, youth groups, and neighborhood associations, played important roles.

Compared to the employment-based health insurance of the first phase, in which employers were in charge of covering the insurance premium, National Health Insurance required the voluntary participation of farmers, fishermen, or small-scale merchants to contribute to paying insurance premiums. This participation was facilitated by the Industrial Associations, which acted as the insurance union at the beginning. In addition, the postal life insurance system established in 1916 could have contributed to gaining the trust of the government-led insurance scheme among the population. Postal life insurance was a nationwide network linking post offices scattered throughout the country, even in remote rural villages. In 1942, the number of subscribers reached 54,546,156 [56], 75% of the total population. This shows that the idea and behaviour of paying insurance premiums in preparation for risk were already deeply rooted in people's minds. However, the revision of the National Health Insurance Law in 1942 made the establishment of the National Health Insurance Association compulsory, and the subsequent rapid increase in membership was undoubtedly due to the enforceability of this law. Since 1940, national treasury contributions to health insurance have increased. Also, dependent benefits started to appear in the statistics from that year and increased significantly until 1945.

It is often mentioned that the occupation policy of the GHQ after the war had an impact on Japan's health policy, but it seems that the impact on post-war universal health coverage was limited or worked in the opposite direction. In his memoirs, Crawford F. Sams, former director of the GHQ Public Health and Welfare Bureau, wrote, "Social security can ruin a nation" and "We did not like the idea of state medicine" [57]. Although GHQ ensured that the freedom of choice of doctors and hospitals was guaranteed, the health insurance system did not undergo significant structural change during the GHQ period and was passed on to the national government after the restoration of sovereignty.

In 1957, the "National Health Insurance Four-Year Plan" was formulated. Four years later, in April 1961, the national health insurance system was implemented in all prefectures as planned, achieving UHC. The four-year plan was most effective in urban areas such as Tokyo and Osaka, but some prefectures such as Iwate, Yamagata, Saitama, Niigata, or Shiga had already achieved universal coverage even before the four-year plan. The pre-war experience with health insurance coverage played an important role in that.

2.6 Onset of Ageing

Koki, literally meaning "rare since ancient times", is one of the life benchmarks to celebrate old age in historical Japan, designating 70 years old. This age also corresponds to the age when Mrs. Orin, the main character of "Narayama bushikō", a fiction based on the folk tale of abandoning the elderly, forced herself to stay in the remote mountain to perish. Throughout Japan's history, old age was celebrated as a rare event and disdained in times of resource constraints. Nationwide statistics on population by age became uniformly available from 1884, and the proportion of the older persons, defined as those aged 65 years and older, remained stable at around 5% until the late 1950s (Fig. 2.11).

However, the stable age structure began to change from around the 1960s due to a drastic decline in both fertility and mortality. The common definition of the speed of ageing is the number of years the proportion of those aged 65 years and over rises from 7% to 14% [58]. This proportion became 7% in 1970 and 14% in 1994 in Japan, making Japan the quickest ageing country in the twentieth century. The speed of ageing was only 24 years, much quicker than France's 115 years, the UK's 46 years, and the US's 72 years.

Population ageing was well recognized among Japanese demographers even in the 1950s. The term "population ageing" was first mentioned in 1955 in the *Journal of Population Problems* [59], around the time when the United Nations released a report on population ageing [60]. In the domain of health, the focus on ageing came

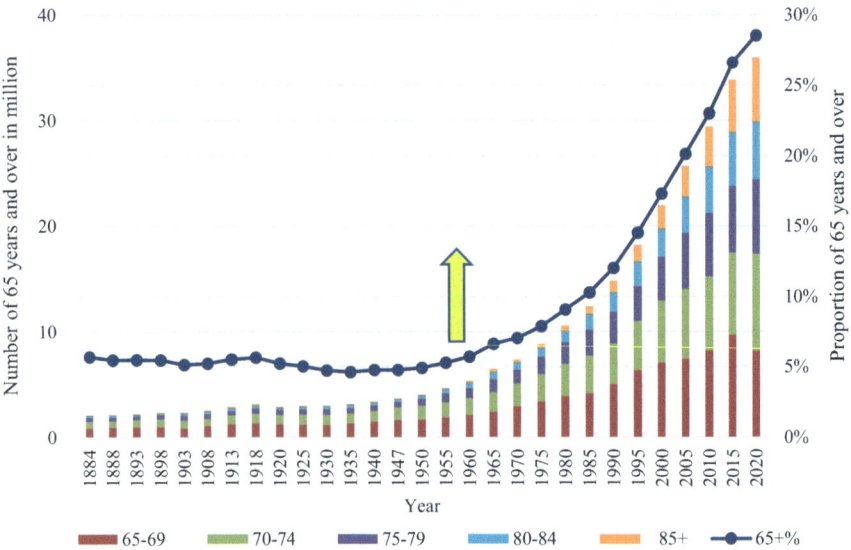

Fig. 2.11 Number and proportion of older persons since 1884 to present. Sources: Historical Statistics of Japan (Japan Statistical Association) from 1884 to 1918, Population Census (Statistics Bureau) from 1920 to 2020

a bit later in the 1960s. The successful control over tuberculosis and the reduction of fertility and infant mortality re-oriented the focus of public health policy from infectious diseases and maternal health to chronic diseases. Universal health and pension coverage was achieved in 1961, and the next aim was to establish a welfare system for the older persons.

In 1962, the ruling Liberal Democratic Party won the election with a manifesto that included promoting the welfare legal system for the older persons. As a result, the Act on Social Welfare for the Elderly came into effect quickly in August 1963. This Act is the first of its kind in the world, although it was the last among the three pillars of the welfare act of Japan, after the Child Welfare Act in 1947 and the Act on Welfare of Physically Disabled Persons in 1949. The Act started at the right moment also from the fiscal point of view. The proportion of Social Benefit, the spending on social security, remained at a low level of 4.28% of GDP primarily due to high economic growth and tax revenue increase [61]. Increasing demand for social welfare for older persons coincided with a high economic boom, and the health and welfare system for the older persons had a promising start.

The Act on Social Welfare for the Elderly aimed to maintain the health and stabilize the living of older persons who are to be loved and respected as they have contributed to the development of society and possess knowledge and experiences gained through long life. The Act strengthened existing welfare measures, such as increasing the number and type of facilities for older persons, but also legalized and capitalized on previous local and voluntary activities, such as yearly medical check-ups, provision of family helpers, supporting senior citizens clubs, or operating welfare centers for the older persons. September 15th was designated as the Day for the Older Persons, and centenarians were celebrated with a congratulatory remark and a silver cup offered by the Prime Minister. The number was small; the first centenarians celebrated in 1963 counted 154.

This Act transferred the welfare of the older persons from mere poverty alleviation through the public assistance scheme, to a wider range of measures to support the life of the older persons. Existing facilities for older persons were mandated to accommodate the poor and lone elderly, but with the new Act, additional types of facility were created, which specialized in offering long-term care in correspondence to the rising needs.

The bedridden elderly started to be noticed as a social problem. In 1968, the Japan National Council of Social Welfare conducted a monitoring survey mobilizing a nationwide network of 130,000 Welfare Volunteers and counted 191,352 bedridden older persons, which corresponded to 5.2% of older persons aged 70 years and over throughout the country [62]. It is best described in the novel "The Twilight Years" by Sawako Ariyoshi, published in 1972, how the ageing care issue crept deeply into Japanese society. The heroine, Akiko, who works in a small law firm in Tokyo, finds out that 3 out of 4 fellow workers in the office face the burden of caring for demented older family members. Caring for (grand) parents or parents-in-law had been a private matter for which each family should take full responsibility. However, around the time of this million-seller novel, the issue surpassed the private realm into to the social sphere. The burden of care surpassed what the family could offer.

Under these circumstances, free medical care for the elderly aged 70 years and over was institutionalized throughout the country by the revised Act on Social Welfare for the Elderly in 1973. This policy was not a nationally led top-down decision but rather a product made out of the struggle between cost-conscious national bureaucrats and motivated local governments. Free medical care started first in Sawamura village (present Nishiwaga town) of Iwate prefecture in 1960, followed by Tokyo prefecture in 1969, then spread to all but two prefectures by 1972. The national government then institutionalized it at the national level. This is a frequent pattern of policy formation in Japan.

Free medical care for older persons suddenly shifted long-term care at home to the medical sphere. As it was free to leave the bedridden persons at the hospital, it was convenient both for families and hospitals. Families were tired of the hard care load, and they did not wish to send their parents to facilities, which were largely stigmatized at the time and also lacked its capacity, so they sent the parents to the hospital. The hospital accepted them and provided care to bedridden older persons, which created a source of income, with payment from public health insurance. The number of these "elderly hospitals" rose very quickly, and so did the number of deaths that occurred in the hospital (Fig. 2.12). In 1976, the number of deaths in hospitals exceeded the number of deaths at home. Dying became more medicalized, along with population ageing and the development of policies.

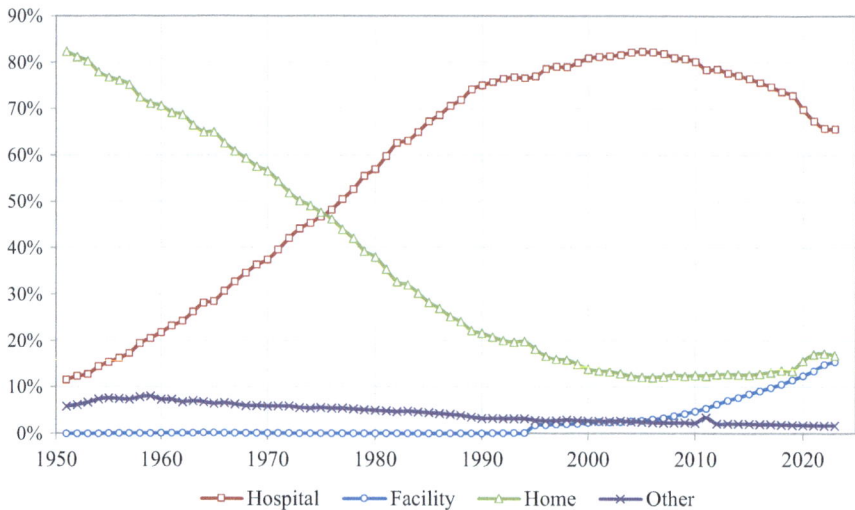

Fig. 2.12 The location of death in Japan, 1951–2023. Source: Vital Statistics (Ministry of Health Labour and Welfare)

2.7 Reforms, Consumption Tax and the Welfare Bubble

The rising number of elderly hospitals, and the surging number of older persons visiting clinics and hospitals for free care led to a sharp increase in health expenditures. As a result, the elderly free medical care policy, which had been in place for less than ten years, was abolished. It was replaced by the Health and Medical Services Act for the Aged, which was enacted in August 1982 and became effective in February 1983. This new law introduced a co-payment system for older persons. The elderly free medical care policy was a mere euphoria during the short-lived high economic growth period, but it also helped shifting the mindset of older persons toward medical care. Previously, they had viewed seeking medical attention as shameful, believing that illness was an unavoidable part of old age. With the free medical care policy, however, they began to feel entitled to receive care.

At the time, a major concern was the disparity among different health insurance funds. When employees retired, they moved to the National Health Insurance, which was maintained by municipalities. This change significantly burdened the National Health Insurance's finances. The 1983 Health and Medical Services Act for the Aged addressed this by introducing a mechanism to balance funds among the various health insurance providers. Another significant reform came with the 1984 revision of the Health Insurance Act, which introduced a co-payment for the insured employee and established the Retiree Health System. This system required employers to cover the healthcare costs of their retirees, while allowing them to use the co-payments of their employees, who did not pay any co-payment prior to that. This system helped the National Health Insurance and also the government, which injected large amount of subsidy to the National Health Insurance. The revision was passed in just one session of the National Diet, as a harsh debate had already taken place during the general election in December 1983 [25]. The Minister of Health and Welfare at the time recalled fighting the election to defend the revision against attacks from left-wing parties, arguing that it was essential for Japan's future [63]. He passed the election, and so did the ruling Liberal Democratic Party, narrowly. The Act was revised shortly after that.

During this period, Japan had to confront unprecedented and unexpected population ageing. While population ageing was viewed favorably in the early 1950s, perceptions shifted dramatically by the late 1970s. Phrases in official government documents, such as the White Paper on Health and Welfare, began to reflect this new reality, with common introductory statements like "the ageing of our country is proceeding without any similar examples in other countries" [64], and "Japan will be the most aged country in the world" [65]. Indeed, this prediction came true.

The unforeseen speed of population ageing is well-reflected in population projections made during the period. Until the 1981 projection, it was assumed that life expectancy at birth would eventually stop increasing once it reached a certain level [66]. This assumption, however, proved to be wrong, and life expectancy continued to increase without limit. On the other hand, fertility—another factor in population ageing—unexpectedly dropped below the replacement level starting in 1974. Demographers had typically assumed that fertility would eventually converge back to the replacement level [67], but this assumption also turned out to be incorrect. With the total fertility rate continuing to sink until it bottomed out in 2005 and life expectancy steadily rising, the resulting proportion of older persons was always higher than projected.

This significant gap between projections and reality exposed the Institute of Population Problem (IPP), which was responsible for Japan's population projections and attached to the Ministry of Health and Welfare, to harsh criticism from the media [68]. This led fellow researchers to create their own projections [69–71] (NUPRI 1982). The projection gap was caused by the unexpected trends of both fertility and mortality. However, the criticism interestingly focused on the fertility decline, not the mortality improvement. The decline in mortality and improvement of life expectancy was universally seen as a positive development, not something to be blamed.

To ease the huge deficit created by free medical care for the elderly, a consumption tax was proposed and finally introduced in 1989. As stated in the first article of the Consumption Tax Act, this tax was intended to be used in part for social security—including pensions, medical care, and long-term care—as well as for measures to alleviate low fertility. Although the introduction of the tax did not immediately raise the country's tax revenue due to an economic slowdown, efforts to improve social security continued.

In 1989, the Gold Plan, a ten-year strategy to promote health care and welfare for the aged, was launched jointly by the Ministry of Finance, the Ministry of Health and Welfare and the Ministry of Internal Affairs. It was a comprehensive welfare package backed by a national government expenditure of 6 trillion yen from 1990 to 1999. The plan included seven key elements: (1) support for home care by increasing the number of home helpers, short-stay capacity, and day service centers, (2) a campaign to "reduce bedridden aged people to zero", (3) the establishment of the "Longevity Social Welfare Fund", (4) the construction of long-term care facilities, (5) the promotion of productive ageing; (6) the promotion of research, and (7) the development of Comprehensive Welfare Institutions for the Aged [72]. Before the end of its initial ten-year period, the New Gold Plan was launched in 1995, which eventually led to the launch of a long-term care insurance system in 2000.

2.8 Toward the Limit of Human Longevity

A significant amount of literature is already available on Japan's long-term care insurance system—its inception, planning, execution, and reforms—in both Japanese and English [73–75], so it will not be discussed here. The basic policy cycle for this system, like others, involves creating, implementing, and revising based on evidence. Life expectancy in Japan has continued to increase until the first year of COVID-19 pandemic. The number of people who have reached 100 years old continues to grow, with 47,107 centenarians in 2023, 300 times more than in 1963. This surge even necessitated a change in tradition, with the prime minister offering silver-plated cups instead of pure silver ones. In the 1970s, as population ageing began, Japan could only look to Western countries for examples to follow. Now, however, Japan is facing an unprecedented demographic situation for the human race and must find its own solutions.

COVID-19 hit Japan just as it did other countries. In 2020, the first year of the pandemic, mortality declined, similar to other Asian countries [76]. This was mainly due to a decline in deaths from pneumonia and influenza, likely achieved through behavioural changes like wearing masks. However, the mortality has worsened in 2022 and has not yet returned to the pre-COVID-19 level by the end of 2024. While COVID-19 has become a common disease and continues to cause many deaths, people have grown accustomed to it, and the fact that thousands of people die from it each month is often ignored. On the other hand, senility deaths are increasing even after adjusting for age structure. It is possible that COVID-19 has changed how people approach terminal care. Just as the Spanish Influenza coincided with the onset of a decline in mortality 100 years prior, COVID-19 could also be a turning point for mortality and healthcare. A detailed analysis should be conducted as the situation progresses.

References

1. Vallin, Jacques. 2011. Politiques de population. In *Dictionnaire de démographie et des sciences de la population*, ed. France Meslé et al., 337–339. UNESCO.
2. Fujikawa, Yu. 1904. *Nihon Igaku Shi (History of medicine in Japan)*. Shokabo. https://dl.ndl.go.jp/pid/833360.
3. Sanitary Bureau of the Home Department. 1877. The first and second report of sanitary bureau. https://dl.ndl.go.jp/pid/1939073
4. Nagayo, Sensai. 1902. *Shokō Shishi (Autobiography of Sensai Nagayo)*. Tōyōbunko 386, Heibonsha. https://dl.ndl.go.jp/pid/1370651/1/4
5. Farr, William. 1885. *Vital statistics*. Offices of the Sanitary Institute.
6. Quetelet, A. 1873. *Congrès international de statistique*. F. Hayez, Imprimeur de l'Académie Royale de Belgique. https://gallica.bnf.fr/ark:/12148/bpt6k6540041f
7. Hasegawa, Sotsusuke. 1959. Meiji shokino shiin bunrui (Cause of death classification at the beginning of Meiji era). *Kōsei no shihyō (Journal of Health and Welfare Statistics)* 6 (11): 24–28.
8. Sanitary Bureau of the Home Department. 1906. *Annual report of sanitary bureau, Meiji 35*. https://dl.ndl.go.jp/pid/836665
9. Bertillon, Jacques. 1903. *Nomenclatures des maladies*. Montévrain: Imprimerie Typographique de l'Ecole d'Alembert.
10. Statistics Bureau of the Cabinet. 1903. Shibō Genin Ruibetsu Chōsa Hōkokusho (Report of the survey on cause of death classification). https://dl.ndl.go.jp/pid/805886
11. Jannetta, Ann Bowman. 1987. *Epidemics and mortality in early modern Japan*. Princeton, NJ: Princeton University Press.
12. Yamamoto, Shunichi. 1982. *Nihon Korera Shi (History of cholera in Japan)*. The University of Tokyo Press.

13. Sugiyama, Shinya. 2001. Ekibyō to Jinkō: Bakumatsu Ishinki no Nihon. In *Reikishi Jinkōgaku no Furontia (The frontier of historical demography)*, ed. Hayami et al. Toyo Keizai.
14. Yamaguchi Prefectural Archives. 2016. *Korori no Kyōfu - Bunsei Gonen no Ryūkō (The terror of "Korori": The epidemics of Bunsei 5)*. The 5th Material Small Exhibition in Fiscal Year Heisei 28. http://archives.pref.yamaguchi.lg.jp/user_data/upload/File/smallexhibition/H28-05.pdf
15. Japan Statistical Association. 1988. *Historical statistics of Japan*. Supervised by Statistics Bureau
16. Oyama, Takuaki. 2010. Kindai Nihon no "Bōeki no Seidoka" ni oite Korera Ryūkō ha "Kettei Jikken" to Narieta ka? (Could epidemic of cholera be the "decisive experiment" for the "institutionalization of epidemic control" in modern Japan?). *Nihon Ishigaku Zasshi (Journal of the Japanese Society for the History of Medicine)* 68(4).
17. Sanitary Bureau of the Home Department. 1885. The ninth annual report of sanitary bureau. https://dl.ndl.go.jp/info:ndljp/pid/836658
18. Shimao, Tadao. 2003. Report and information: Mortality rate of pulmonary TB in the late 19th century. *Kekkaku* 78 (10): 633–636.
19. Suzuki, Takao. 2014. *Hone kara Mita Nihonjin: Kobyōrigaku ga Kataru Rekishi (Japanese seen from bones - History through paleopathology)*. Kodansha.
20. Tanba, Yasuyori. 984 Ishinpō. e-Museum, National Treasures & important cultural properties of National Institutes for Cultural Heritage, Japan.
21. Iwasaki, Tatsuro. 1981. Nihonni okeru kekkaku no rekishi - Kekkaku wa Yōroppajin ga denpa shitanoka (History of tuberculosis in Japan - Did Europeans transmitted the tuberculosis). *Kekkaku (Tuberculosis)* 56 (8): 407–422.
22. Honma, Gencho, edited by Chokyu Kimura. 1864 (1935). Naika Hiroku. Nihon Kanpō Igakukai Shuppanbu. https://dl.ndl.go.jp/info:ndljp/pid/1050469
23. Suda, Keizō. 1973. *Hida O Jiin Kakochō no Kenkyū (Research on the Hida O Temple necrology)*. Iryohojin Seijinkai Sudabyoin.
24. JATA: Japan Anti-Tuberculosis Association. 2016. *Shogen de Tsuzuru Kekkaku Taisaku = Koshu Eisei no Rekishi (Testimonies of tuberculosis control = A history of public health)*.
25. Ministry of Health and Welfare Editorial Committee for 50 years History of Ministry of Health and Welfare. 1988. *Kōseishō 50 nen shi (50 years history of Ministry of Health and Welfare)*. Foundation-Institute for Research of Health and Welfare Problems.
26. Ishihara, Osamu. 1914. *Eiseigakujō yori mitaru jokō no genkyō*. Kokka Igaku Kai. https://dl.ndl.go.jp/info:ndljp/pid/946989
27. Aoki, Masakazu. 2003. *Kekkaku no Rekishi (History of tuberculosis)*. Kodansha.
28. Sanitary Bureau of the Home Department. 1922. *Ryūkōsei Kanbō (Epidemic cold)*. https://dl.ndl.go.jp/info:ndljp/pid/1148597
29. Hayami, Akira. 2006. *Nihon o osotta Spain influenza - Jinrui to uirusu no Daiichiji Sekai Sensō (Spain influenza which attacked Japan - The first world war of human and virus)*. Fujiwara Shoten.
30. Sanitary Bureau of the Home Department. 1920. *Annual report of sanitary bureau, Taishō 7*. https://dl.ndl.go.jp/pid/985011
31. Sanitary Bureau of the Home Department. 1922. *Annual report of sanitary bureau, Taishō 9*. https://dl.ndl.go.jp/pid/836665
32. Matsuura, Kōichi. 1958. Reformation of Japanese pre-census life tables. *Igaku Kenkyu* 28 (7): 138–153.
33. Mizushima, Haruo. 1963. *Seimeihyō no Kenkyū (Research on life table)*. The Research Institute of Life Insurance Welfare.
34. Okazaki, Yoichi. 1986. Population of Japan in the Meiji-Taisho era - Re-estimation. *Journal of Population Problems (Jinkō Mondai Kenkyū)* 178:1–17. https://www.ipss.go.jp/syoushika/bunken/data/pdf/14167501.pdf.
35. Yasukawa, Masaaki. 1977. *Meiji Taishō nenkan no jinkō suikei to jinkō dōtai. Jinkō no Keizaigaku (Economicis of population)*, 149–189. Shunjusha.

36. Sanitary Bureau of the Home Department. 1912. *Annual report of sanitary bureau, Meiji 43.* https://dl.ndl.go.jp/pid/836670
37. Morita, Yūzo. 1944. *Jinkō Zōka no Bunseki (Analysis on population increase).* Nippon Hyoron Sha. https://dl.ndl.go.jp/pid/1459595.
38. Takase, Masato. 1991. Population, birth and death in Japan for the period 1890-1920. *The Journal of Population Studies (Jinkōgaku Kenkyū)* 14:21–34. https://doi.org/10.24454/jps.14.0_21.
39. Nishida, Shigeki. 1996. The contribution of medical techniques to the decline of infant mortality in Japan. *Bulletin of National Institute of Public Health* 45 (3): 292–303.
40. Health and Hygiene Investigation Commission. 1917. The 1st report of the health and hygiene investigation commission. https://dl.ndl.go.jp/pid/1152476
41. Hiroshima, Kiyoshi. 1980. Essay on the history of population policy in modern Japan - Around the concept of "population quality". *The Journal of Population Problems (Jinkō Mondai Kenkyū)* 154:46–61.
42. Homei, Aya. 2023. *Science for governing Japan's population.* Cambridge University Press. https://www.cambridge.org/core/books/science-for-governing-japans-population/E7A44D3331E27F04452A5C42CCF2354C.
43. Ito, Shigeru. 1998. The decline in infant mortality in pre-war Japan. *Socio-Economic History (Shakai-Keizai-Shigau)* 63 (6): 1–28.
44. Saito, Osamu. 2008. Infant mortality and Aiiku Village schemes in prewar Japan. *Socio-Economic History* 73 (6): 33–55.
45. Higami, Emiko. 2016. *Kindai Osaka no nyujishibō to shakaijigyō (Infant mortality and social work in modern Osaka).* Osaka University Press.
46. Takemura, Kōtaro. 2003. *Nihon Bunmei no Nazo wo Toku - 21 Seiki wo Kangaeru Hinto (Unraveling the mysteries of Japanese civilization: Hints for thinking about the 21st century).* Rabat: Seiryu Publishing.
47. Japan Water Works Association. 1967. Nihon Suido Shi (History of water works in Japan).
48. Maruyama, Hiroshi. 1953 (reproduced in 1989). Shiji wo Shite Sakebashimeyo (Let the dead child cry out). *Nature* 8:2, 4, 6, 8, 12; Nōsan Gyoson Bunka Kyōkai.
49. Higami, Emiko. 2013. Infant mortality and industrialization in the Osaka City between the world wars. *Historia, The Osaka Historical Association* 236:131–151.
50. Nishida, Shigeki, Masabumi Kimura, and Kenji Hayashi. 1987. Marriage rate, divorce rate and birth rate in Japan for the period between 1872 and 1898: A re-examination. *Japanese Journal of Health and Human Ecology* 53 (4): 184–191. https://doi.org/10.3861/jshhe.53.184.
51. Ota, Tenrei. 1976. *Nihon sanji chōsetsu hyakunenshi - Meiji Taishō Shōwa shoki made (100 years history of birth control in Japan - From the Meiji, Taisho and early Showa periods).* Shuppan Kagaku Sōgō Kenkyusho.
52. Mainichi Shinbun Sha Jinkō Mondai Chōsakai ed. 2000. *Nihon no Jinkō - Sengo 50 nen no Kiseki (Population of Japan - The post-war 50 years' trajectory).*
53. Ministry of Health and Welfare. 1976. Isei Hyakunen shi Shiryōhen (100 years of medical act, reference part). https://dl.ndl.go.jp/pid/12012111
54. IPSS: National Institute of Population and Social Security Research. 1958–2019. Annual report on social security statistics. https://www.ipss.go.jp/site-ad/index_Japanese/securityAnnualReport.html
55. Takagi, Yasuo. 1994. Kokumin kenkō hoken to chiiki fukushi (National Health Insurance and community welfare). *Quarterly of Social Security Research* 30 (3): 249–260. https://www.ipss.go.jp/syoushika/bunken/data/pdf/sh300305.pdf.
56. Insurance Agency. 1942. Hoken Chōsa Ihō (Insurance Research Report). No. 51.
57. Sams, Crawford F., edited by Zabelle Zakarian. 1998 (2015). *Medic: The mission of an American military doctor in occupied Japan and Wartorn Korea.* Taylor & Francis, Routledge.
58. Kinsella, Kevin, and Wan He. 2009. *An aging world: 2008.* Washington, DC: U.S. Census Bureau, International Population Reports, P95/09-1, U.S. Government Printing Office. https://www.census.gov/library/publications/2009/demo/p95-09-1.html

59. Kuroda, Toshio. 1955. Kōnenka Genshō no Jinkōgaku teki Kenkyū 1 (Demographic research on ageing phenomena 1). *Journal of Population Problems (Jinkō Mondai Kenkyū)* 61:8–62. https://www.ipss.go.jp/syoushika/bunken/data/pdf/14206703.pdf.
60. United Nations. Department of Economic and Social Affairs. 1956. The aging of populations and its economic and social implications. ST/SOA/Series A/26.
61. IPSS: National Institute of Population and Social Security Research. 2024. Financial statistics of social security in Japan 2022.
62. Japan National Council of Social Welfare. 1968. Kyotaku Netakiri Rōjin Jittai Chōsa Houkokusho (Report on the survey of bedridden elderly people living at home).
63. Hayashi, Yoshiro. 1984. *Kōsei gyōsei to watashi (Health and welfare administration and I)*. Gendai Keizai Kenkyūkai. https://dl.ndl.go.jp/pid/1200763
64. Ministry of Health and Welfare. 1970. *Kosei Hakusho (Health and welfare white paper) Showa 45*. Ministry of Finance Printing Bureau https://www.mhlw.go.jp/toukei_hakusho/hakusho/kousei/1970/
65. Ministry of Health and Welfare. 1995. *Kosei Hakusho (health and welfare white paper) Heisei 7*. https://www.mhlw.go.jp/toukei_hakusho/hakusho/kousei/1995/
66. IPP: Institute of Population Problems, Ministry of Health and Welfare. 1982. *Future population projections for japan by sex and age for 1980–2080, prepared in November 1981*. Institute of Population Problems, Research Series No. 227, April 1, 1982. https://www.ipss.go.jp/syoushika/bunken/data/pdf/J08409.pdf
67. IPP: Institute of Population Problems, Ministry of Health and Welfare. 1987. *Population projections for Japan: 1985-2085*. Institute of Population Problems, Research Series No. 244, February 1, 1987. https://www.ipss.go.jp/syoushika/bunken/data/pdf/J08429.pdf
68. Ota, Hiroyuki. 2002. Jinkō mondai kenkyūsho to iu mondai |The problem of the institute of population problems. *Asahi Shinbun Weekly ARERA* 2002 (6): 24–27.
69. Kuroda, Toshio. 1980. *Nihon no shōrai jinkō ni tsuiteno nichidai suikei (Population projection of Japan by Nihon University)*. Nihon University.
70. Yasukawa, Masaaki. 1979. *Waga kuni no shorai jinko suikei - Showa 53 nen Yasukawa suikei (Future population projection of Japan - Showa 53 Yasukawa projection)*. 47th Japan Statistical Society Research Report Material, Kagawa University, S.54.7.25. https://doi.org/10.11329/jjss1970.10.68
71. Fujimasa, Iwao, and Toshiyuki Furukawa. 2000. *Uerukamu jinkō genshō shakai (Welcome, the society of population decline)*. Bungeishunju: Bunshun Shinsho 134.
72. Foundation of Social Development for Senior Citizens. 1995. Ten-year strategy to promote health care and welfare for the aged (Ten-year gold plan for the welfare of the aged). Editorial Supervision: Health and Welfare Bureau for the Elderly, Ministry of Health and Welfare of Japan.
73. Campbell, John Creighton. 2014. Japan's long-term care insurance program as a model for middle-income nations. In *Universal health coverage for inclusive and sustainable development: Lessons from Japan*. World Bank Studies, ed. Naoki Ikegami, 57–67. Washington, DC: World Bank. https://doi.org/10.1596/978-1-4648-0408-3. License: Creative Commons Attribution CC BY 3.0 IGO.
74. Tamiya, Nanako, et al. 2011. Population ageing and wellbeing: Lessons from Japan's long-term care insurance policy. *Lancet* 378.
75. UNESCAP. 2015. Long-term care of older persons in Japan. SDD-SPPS project working papers series: Long-term care for older persons in Asia and the Pacific. http://www.unescap.org/resources/long-term-care-older-persons-japan
76. Hayashi, Reiko. 2022. COVID-19 and mortality decline in Asia in 2020. *Journal of Population Problems (Jinkō Mondai Kenkyū)* 78 (4): 493–508. https://doi.org/10.50870/00000436.
77. WHO: World Health Organization. 1995. District Health Systems: Global and Regional Review Based on Experience in Various Countries. *Division of Strengthenng of Health Services*. WHO/SHS/DHS/95.1.
78. NUPRI: Nihon University Population Research Institute. 1982. Nihon Daigau Jinkō Kenkyusho Jinkō Suikei: Jinkō keizai moderu ni motoduku 21 seiki eno tenbō (Population projection of NUPRI: Outlook for the 21st century based on population and economic models). Tokyo.

Chapter 3
Migration: The Most Effective Population Policy?

3.1 Introduction

Migration is one of the components which determines population, thus the migration policy is one of the population policies. Indeed, the Commission for the Investigation of Problems of Population and Food (CIPPF), the first national commission on population established in 1927, clearly stipulated emigration policy as the first measure to ease the population pressure, rather than the fertility control [1]. The mobility control policies in the Edo period, such as *Sakoku* (closure of the country) in terms of international migration, population registration and *Sekisho* (control gate) system in terms of internal migration, were replaced by modernized government after the Meiji Restoration with a variety of new mobility policies. Modernization came along with the mobility of people. The opening of the country to the world and the abolition of the *Sekisho* system, which enabled free movement within Japan, were central to this "restoration".

In contrast to fertility or mortality, the measurement of migration is complex. It can be measured by flow or stock, by the international boundaries or internal, domestic, prefectural or municipal boundaries that change time to time, by lifetime or 1 year, 5 years, and so forth [2]. Among many indicators, three indicators are chosen and shown in Fig. 3.1. The inter-prefectural lifetime mobility is an indicator of internal migration, and the number of overseas Japanese and foreign nationals in Japan is the indicator of international migration.

The magnitude of internal migration, defined here as the number of people who are living outside of their birth prefecture, was very low in the nineteenth century, then increased to 15.1% in 1920, and remained almost the same until 1950. However, by the next available data point in 1986, the proportion had jumped to 26.3% in 1986. The post-war economic growth coincided with increased internal mobility. The proportion decreased around the turn of the twenty-first century, but since 2011, it turned to an upward trend. This is partly due to the increasing number of foreign

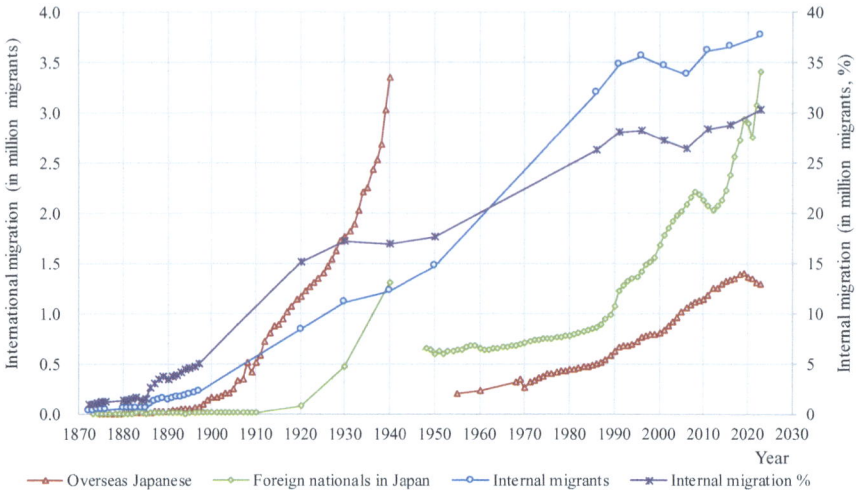

Fig. 3.1 The trend of internal and international migration. Note: Internal migration is defined here as the number of people who are living outside of their birth prefecture. The proportion denominator is the total population. Substantial increase is reported after 1940, notably Japanese in Manchukuo and Koreans in Japan, but it is not shown here as the other portions of statistics are missing. Sources: Overseas Japanese by Survey on the Japanese Nationals Overseas (Ministry of Foreign Affairs), compiled in Hayashi [3]. Foreign nationals in Japan by Statistics on Foreign National Residents (Ministry of Justice) and Japan Statistical Association [4]. Inter-prefectural lifetime mobility from 1872 to 1897 by Family Register, from 1920 to 1950 by the Population Census (Statistics Bureau) and from 1976 to 2018 by the National Survey on Migration (IPSS)

nationals living in Japan. In any case, the number of internal migrants is 37.7 million people, roughly ten times more than the number of foreign nationals of 3.4 million.

As for international migration, the number of overseas Japanese increased from the 1880s to the height of 3.3 million in 1940. It increased further notably by emigration to Manchukuo. The defeat of WWII caused massive return migration to the country, and the number remained low throughout the period of economic growth in the 1960s and 1970s. The gradual increase since then, however, did not attain the pre-war level. Since 2020, the number has decreased due to the COVID-19 pandemic.

The number of foreign nationals living in Japan was counted from the start of Meiji, and it increased modestly from 4190 in 1873 [5] to 54,320 in 1930 [6]. However, if we include nationals of overseas territories living in mainland Japan, the number increases significantly to 1,304,286 in 1940. It increased further notably by the Koreans who kept on moving to Japan. During this period, the Koreans were the most numerous, followed by the Taiwanese. Many of them returned after WWII, and the number was substantially reduced in 1947. Similar to the case of overseas Japanese, the number of foreign nationals in Japan did not grow during the high economic boom in the 1960s and 1970s, but a sharp increase started from the end of the 1980s, reaching 3.4 million in 2023, overcoming the two plunges caused by the Great East Japan Earthquake in 2011 and the COVID-19 started from 2020. The year 2023 is symbolic, as the number of foreign nationals in Japan exceeded the peak of 3.3 million of overseas Japanese in 1940. Clearly, Japan had changed from an emigration to an immigration country. All these trends faithfully reflect the major social and political events of each era. This chapter will explore how these migration trends arose, whether spontaneously or through deliberate policy.

3.2 Pre-Modern Mobility: In the Edo Period

Sakoku, often translated as "locking the country", is a famous term used to describe the border control of Edo period Japan (1603–1868). This word was not the official term used by the Edo government, but a translated word from a Dutch description of Japan observed in two years from 1691 to 1692 [7–9].

In fact, in the Edo period, it was not completely locked, but foreign trade, as well as international mobility, were severely controlled. Among the Westerners, only the Dutch were allowed to trade on the small island called Dejima in Nagasaki. The Chinese were allocated a district to stay in Nagasaki. The Korean relationship and the delegations' visit were managed through Tsushima Han, the island domain of the Edo Government, which lies between Korea and Japan. The trade between mainland Japan and the Ryukyu dynasty, as well as with the Ainu, was controlled through the Satsuma and Matsumae local governments, respectively. Hence, the entrance of non-Japanese was limited to these channels, and the Japanese who moved out from the territory were even fewer, only limited to those such as fishermen who accidentally drifted away by shipwreck or Tsushima's language translator who landed sometimes to the Korean peninsula.

Internally also, the movement of people was controlled. At the time, the *Shumon/Ninbetsu Aratame* (Religion/Person Verification) register, originally introduced to ascertain each person not to be Christian, was employed also for controlling people's movement between villages and cities. The village head who was responsible for keeping the register granted the certificate to his fellow villager, so this villager could use the certificate as a passport for travel and temporal living within Japan. *Sekisho*, the checkpoints were established in the critical nodes of the road network to control the movement of people and trade. There were 53 checkpoints towards Edo city [10].

The purpose of checkpoints was often described as controlling "entering weapons and exiting women" to and from Edo, the capital city. Entering weapons to Edo should be controlled to keep the safety of the capital. The "exiting women" in question were the wives of regional lords, who were required to stay in Edo as de facto hostages. The escape from Edo was considered to be the defiance of the Tokugawa order. However, the control of women was made on both sides, exiting as well as entering the checkpoint from and towards Edo. The control might have been not only for controlling the wives of the regional lords but also for women in general, and it was stricter than for men. Controlling the movement of women—the "source of the population"—could thus also be considered a form of population policy.

These international and internal mobility controlling systems had both merits and demerits. As for the international border control, the original intention to halt the propagation of Christianity and possibly monopolizing international trade helped Edo governance to be stable for 250 years, kept peace and helped the development of Japanese culture [7, 11, 12]. The internal movement control did not necessarily prevent the movement of people, and the controlled movement developed the road network and the total length of roads stretching up to 1500 km [13], connecting cities and towns of the archipelago. Migrating out to seek a job was a common practice, and as much as 50% of men and 62% of women experienced that type of migration during their lives for the cohort born from 1773 to 1825 in Seijo village in Mino country, present-day Gifu prefecture [14]. As for the control made at the checkpoints, it was often possible to bypass them by taking detours, although this was severely punished if discovered. During times of bad weather and poor harvest, some regional lords closed the border so that massive vagabonds would not be able to rush into the territory in search of food. On the other hand, it implies that during normal conditions, beggars managed to enter the wealthier cities and towns in search of food. The internal mobility control thus differed from time to time, from place to place.

The international migration control was an Edo invention in parallel with the same policies taken in China and Korea, and it was rather the exception throughout the history of Japan. Japanese basic governance was brought in by the Chinese in the seventh to ninth century, and along with the global Era of Voyage starting from the fifteenth century, many Japanese migrated out to various places also. There were 3000 Japanese settlers in Manila in 1620, and similar Japanese settlements were created in Ayuthaya, Hoian, Patani or Phnom Penh in South East Asia but also in Lima, Peru, where 20 Japanese settled and Mexico City around 1610s [15]. In addition, various officials, Japanese Christian missionaries and travellers moved internationally in the 1610s, notably Tanaka and Hasekura delegations composed of 22 and 180 members, respectively, which visited Mexico and Europe. After the "locking policy" with the orders issued in the 1630s, the Japanese who went abroad were very limited, only the illegal students or castaways. However, their records are well known. Souha Hatano (the first) stayed 3 or 5 years in the Netherlands in the 1650s or 1660s to study medicine and eventually returned to Japan and opened a surgery clinic in Osaka. Kodayu Daikokuya drifted away on his sea trade route from Ise to Edo, landed on Amchitka Island in 1782 and returned to Japan in 1792 after staying in Saint Petersburg and meeting Catherine the Great. Tudayu, and three other crew of Wakamiya-maru shipwrecked in 1793 on the way from Ishinomaki to Edo, travelled around the world together with Mr. Rezanov, the Russian diplomat, onboard the ship named Nadezhda and came back to Nagasaki in 1804. Nakahama Manjiro boarded on a fishing boat as a cook, lost in the sea and found by the American whaler, stayed and learned in the United States before returning back to Japan in 1851. Although the number was very limited, those people who moved across the border made a significant impact on Japanese society. However, overall, the closure of the country did not facilitate the emergence of free spirit such as in eighteenth century France [16], and caused population stagnation recorded between 1732 to 1842.

3.3 Meiji Restoration as the Liberalization of Mobility

The Meiji Restoration was a political coup d'état but also the liberation of mobility of people. In 1869, the *Sekisho* (gate control) system was abolished. The Family Register Act was put in place in 1872 and replaced the village-based registration that existed so far. By this Act, the travellers and those living more than 90 days outside their originally registered place were ordered to register their temporary stay [17, 18].

At the beginning of 1872, there were 335,886 in-migrants and 184,099 out-migrants who crossed the prefectural border registered throughout the country. The difference, 151,787, is due to the registration practice; people tended to register at the destination but not as much at the origin. The proportion of in-migrant to the registered population, 33,087,390, was only 1.0%. That year, the first railway was launched from Yokohama to Shinbashi for an operational distance of 29 km. At that time, people travelled primarily on foot, and their mobility was low. The national level statistics, available from 1872 to 1897, show a slow increase until 1883, followed by a short decrease, then a rapid increase, both in number and proportion (Fig. 3.2).

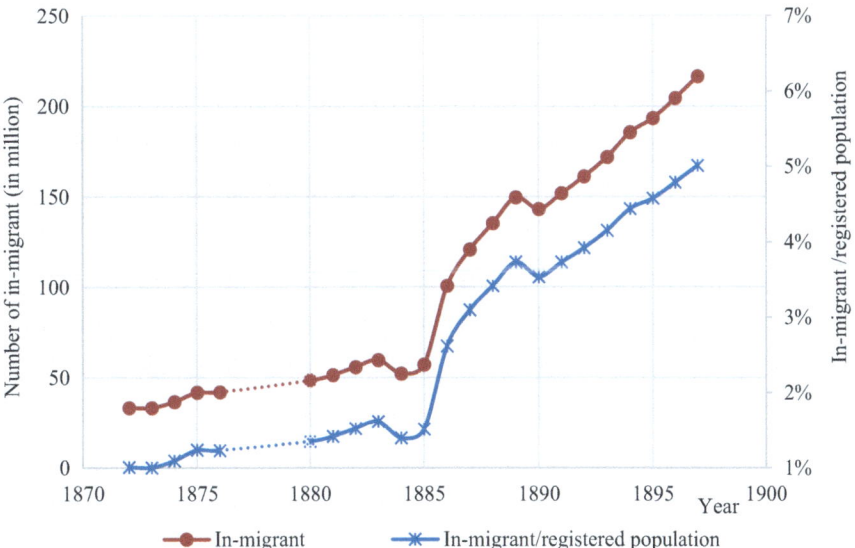

Fig. 3.2 The trend of internal migration, 1872–1897. Source: Statistics Bureau [19]

The time of the decline of internal migration in 1884 and 1885 corresponds to the governmental budget cut following the deflationary policy introduced by Masayoshi Matsukata, then the finance minister. On the other hand, the "unprecedented windstorm" [20] hit Japan in August and September 1884. Probably affected by both, the number of railway passengers declined in 1884 and 1885 [4]. As for the migration decline in 1890, similar economic or climate disasters were not on record, but it can be due to the first national general election held that year. The family register-based population was used to determine the seats of parliamentarians; some sort of cleaning of the family register might have been conducted.

It might be assumed that pandemics of fatal diseases would prevent people from moving. However, the years when the massive Cholera outbreaks occurred in 1879 and 1886, killing 105,786 and 109,012 people, respectively, did not affect the increasing trend of migration.

The same statistics based on family register are available by prefecture for the same period from 1872 to 1897 (Fig. 3.3). The largest number of in-migrants who crossed the prefectural boundary throughout the period was recorded in Tokyo prefecture, followed by Osaka and Hokkaido. The movement of people to the capital city and the second biggest city progressed since the earliest years of the Meiji era. However, the third destination, Hokkaido, explains that mobility was driven not only by the natural urbanization process but also by policy.

There were a few attempts of peopling of Hokkaido during the Edo period but the Meiji government immigration policy was considered to be the first governmental attempt in Japanese history [21]. The Development Bureau (*Kaitaku-shi*) was established in 1869, and military-settler colonists (*Tondenhei*) were institutionalized in

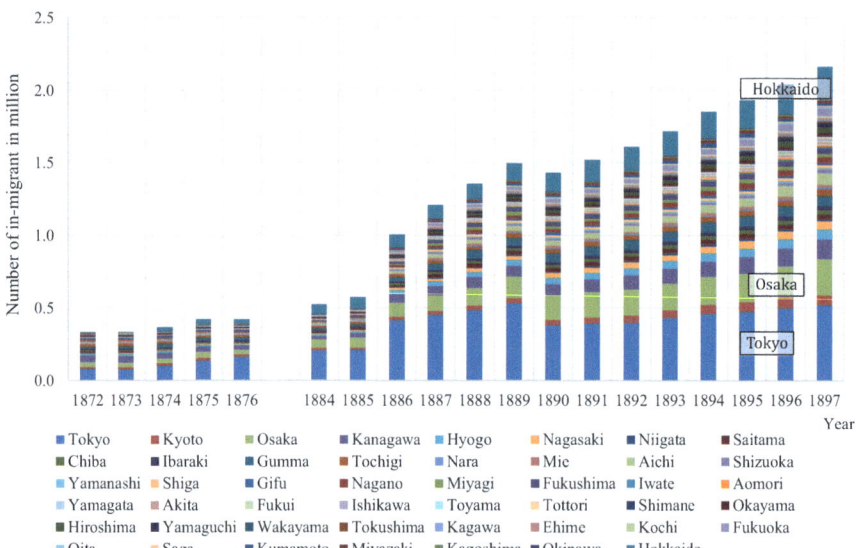

Fig. 3.3 Number of in-migrants by prefecture. Source: Statistics Bureau [19]

1874. The purpose was to protect the territory and develop the land, and it helped create jobs for the former Samurai who were left in poverty due to the abolishment of the old regime [22]. Also, the low population density of Hokkaido was considered as an opportunity [23], and the motives of the national defence and expansionism was backed by the newly imported Malthusian overpopulation theory [24, 25]. The number of colonists increased, and by the time the institution was abolished in 1904, there were 39,911 colonists in 7337 households in 37 military villages [26]. The land of Hokkaido was well developed by then, and there was no more need to implement a special regime, which bound the military settler-colonists with regulations different from the ordinary residents. The migration of citizens to Hokkaido from the rest of Japan continued until 1953, when the outflow surpassed the inflow [27].

Gunma prefecture, now a typical rural prefecture, was a major migration destination that ranked fifth in most of the years of the early Meiji era. It is improbable that people gathered there without specific incentives. The reason could be the Tomioka Silk Mill, one of the first governmental factories built in 1872 in the prefecture. The national investment for the operation of the factory, as well as sericulture promotion and railway extension can be good reasons for attracting people coming from the other parts of the country.

3.4 Shifting Destinations Amid an Evolving International Climate: Emigration to Ease the Population Pressure

The nineteenth century was the era of international migration of the world. Japan was well integrated with this global trend. The first person who obtained a Japanese passport was Mr. Namigorō Sumidagawa, a circus entertainer leading the Imperial Japanese Troupe who sailed to the US in 1866 [28]. It was a voluntary, business-based trip, not guided by policy. The earliest out-migration of Japanese workers to Hawaii, Guam and California started in 1868, the first year of Meiji. Foreign companies initiated it but were quickly halted by the Japanese government, which feared the slave-like treatment of the countrymen [27]. With the increasing "pull" force from the new world due to the abolition of the African slave trade as well as the development boom of new territories in various parts of the world, the Japanese government responded [29]. The first bilateral immigration convention was ratified between Japan and the Kingdom of Hawaii in 1886, and governmentally protected emigration started. Mr. Robert Walker Irwin, a descendant of Mr. Benjamin Franklin and the first person who registered international marriage in Japan was the signer of the convention as the Consul-General of the Kingdom of Hawaii. The Japanese signer was Mr. Kaoru Inoue, then the foreign minister of Japan, who was also the business partner of Mr. Irwin. It is not certain if these early emigration policies were based on a perception of "over-population"; they may have been motivated more by simple entrepreneurship.

The number of Japanese emigrants rose thereafter. According to the flow statistics, the destination shifted first to Hawaii, secondly to North America, and thirdly to Latin America [30, 31], which the statistics faithfully depict (Fig. 3.4). Hawaii was the closest pulling destination with the rapid development of sugar cane production implemented by international entrepreneurs. The shift to mainland US as well as British Northern America (Canada) was partly due to replacing Chinese who were shut out by those countries [29], but within 40 years, the Japanese followed the same itinerary. The 1882 Chinese Exclusion Act was followed by the 1924 Immigration Act of the US which set a reduced quota for new immigrants, including Japanese. But the new destinations were abundant in Latin America, first to Mexico, then Peru and Brazil.

As for the stock statistics, the Ministry of Foreign Affairs first counted the number of overseas Japanese by counting the number of passports issued and subtracting the number of returned passports. This system changed in 1889 to the registered number at the Japanese Consulates overseas. This practice continues up to present. The number of overseas Japanese fluctuates in some years due to the missing reports due to wars, insurgences of the destination countries but throughout the years one can observe the trend (Fig. 3.5).

At the time of global imperialism, the destinations of overseas Japanese are easy to understand. The statistics on the number of overseas Japanese were classified in the columns of "territory of" Russia, UK, France, US, and so forth. Those territories needed people and accommodated migrant-friendly environments. On the contrary, the number of those who resided in independent states such as Thailand, Iran, or Turkey was small. Before Japan expanded its own colonization, it benefitted from

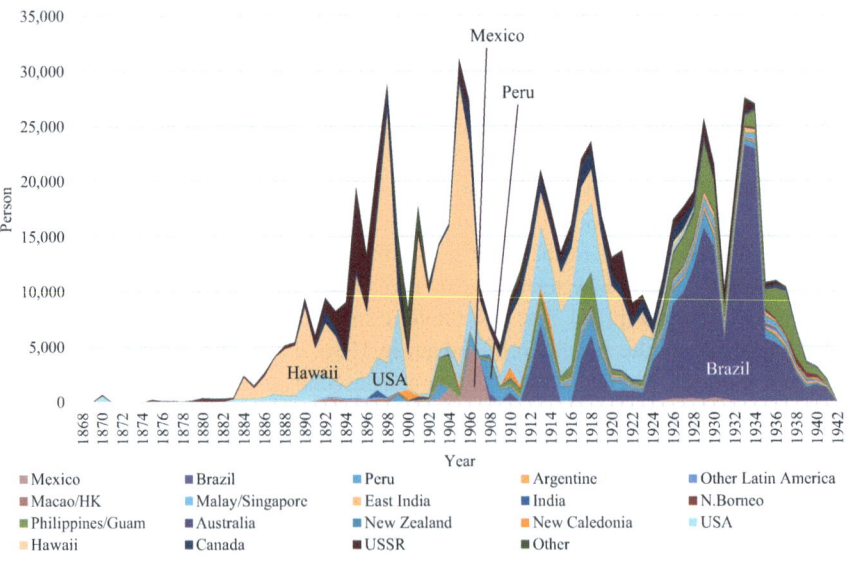

Fig. 3.4 Annual flow of Japanese emigrants by destination. Source: JICA [31]

3.4 Shifting Destinations Amid an Evolving International Climate: Emigration to Ease... 53

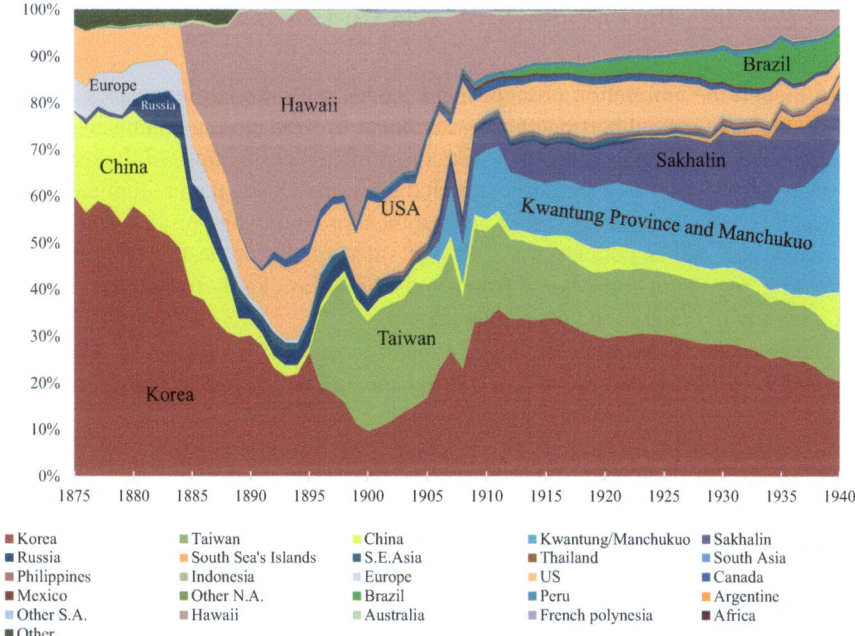

Fig. 3.5 Overseas Japanese population by destination (%, stock). Sources: Japan Statistics Yearbook (Statistics Bureau). Korea, Taiwan, Sakhalin, Kwangtong Province, and South Seas Islands are by respective governments' statistics reports. Parts of the data are compiled by the Japan Statistical Association [4]. Overseas Japanese Population Table (Ministry of Foreign Affairs) for 1933, 1939 and 1940. In: Hayashi [3]

the imperial global network, which created trade, industry and the need for manpower.

The overseas Japanese were male-dominant, especially in earlier periods. The sex ratio was high; for example, 394 in 1900, 80% were male. However small the number, there was a particularity of Japanese female out-migrants. Of the 498 women who were granted passports in 1880, half (233) were with the motif "handwork and slavery", depicted in the first year the emigration statistics appeared in the first volume of the Japan Statistics Yearbook published in 1882 [32]. The most were in China (66), followed by Korea (45) and Russia (33). The so-called *Jōshigun* ("women's army") were women who moved abroad to support themselves through prostitution. These women were originally from Nagasaki, and the destination expanded to different parts of the world, even to Madagascar and Cape Town [29]. Although they were sometimes more efficient than military men in foreign countries [33], the continuous practice of these women until the end of WWII shows the gender reality of Japan at the time; entertainers and prostitutes were tacitly acknowledged occupations and incorporated into the society.

3.5 Peopling the Greater East Asia Co-Prosperity Sphere

Even though the emigration brought many Japanese to the farthest places on the planet, the most crucial portion of overseas Japanese lived close to mainland Japan. Korea, Taiwan, and mainland China were the principal destinations of the overseas Japanese (Fig. 3.5). From the earliest period up to the 1880s, the predominance of Korea and China was mainly due to geographical proximity and historical continuity, when emigration policies were not yet substantially orchestrated. Later, the increasing number of Japanese in Korea and China was closely related to policies, not precisely of emigration but of colonization.

The processes of the early Meiji migration of Japanese to Taiwan and Korea were not similar. For Korea, during the Edo period, the diplomatic relationship was maintained, and 12 missions from Korea to Japan were carried out from 1607 to 1811. There was an official Japanese Pavilion at present-day Busan which was maintained by Tsushima Domain [34]. This Soryo Pavilion was turned into the Japanese Consulate after the Ganghwa Island incident in 1875, in which a Japanese vessel landed on the Korean island and killed 35 Korean soldiers. The number of passports granted to Japanese for going to Korea from the first year of Meiji to 1875 totalled 1965, and the number rose over the years. In 1896, the number of Japanese residing in Korea was 12,571.

As for Taiwan, Japan sent troops in 1874 in retaliation for 54 Japanese decapitated by Taiwan's indigenous people, but until the First Sino-Japanese War, no Japanese were residing on the island [35]. The first statistics of Japanese residents in Taiwan, which appeared in the statistics yearbook of the Government-General of Taiwan, was 10,584 persons in 1896. Considering there were 32,226 Japanese arrivals in one year of 1897, it is probable that in one year of 1896, the Japanese rushed to Taiwan all of a sudden after the Treaty of Shimonoseki was signed in 1895. The high average number of household members, 4.88, would back this sudden settlement as many people stayed in dormitory-type housing just after their arrival.

Although the start was quite different, the number of Japanese in Korea and Taiwan rose quickly with similar speed until it counted around 50,000 in 1905 in both territories. However, since the annexation of Korea in 1910, the number of Japanese in Korea rose more quickly than in Taiwan. There were two times more Japanese in Korea than in Taiwan, 689,747 in Korea and 348,962 in Taiwan in 1940.

The statistics are a faithful witness of Japanese expansion. In addition to Taiwan and Korea, in Kwantong Province, the tip of the Liaodong Peninsula in China containing Dalian and Lushun, the Japanese increased from 1905 after the Russo-Japanese War. In the same year, the southern part of Sakhalin became the Japanese territory under the Portsmouth Peace Treaty, and the population statistics started. The Japanese population of South Sea's Islands started to appear in 1920 statistics, published in December 1921, slightly before Japan was mandated to govern those islands by the League of Nations in 1922.

To rule the area under control, there should also be Japanese people. Not only governmental officers but farmers, merchants and ordinary people were needed.

There were many policies implemented to "plant the people", the literal meaning of "colonization" in Japanese *"Shokumin"*. The emigration policies to farther areas, such as in Hawaii, the US and Latin America, were replaced by the policies for closer areas. Partly because Japanese immigrants faced opposition from the destination authorities but also because there was a notion that sending the countrymen to remote areas was a waste. After the victory of the Russo-Japanese War, Marquess Jutaro Komura, then the Minister of Foreign Affairs, made a remark in the 25th Imperial Parliament in 1909 that it should be avoided that the Japanese people dispersed unnecessarily in remote foreign territories, and they should be driven to Manchuria [30].

The migration policy was one of the pillars of the population policy. In 1927, the first report of the Commission for the Investigation of Problems of Population and Food (CIPPF) addressed the importance of the promotion of migration and the protection of Japanese migrants [1]. On the other hand, the third report released in 1928 advocated that the Japanese migration in these areas should accompany the population development of the local people in Korea, Taiwan, Manchuria-Mongolia, Western Siberia and the South Seas Islands.

These policies of the peacetime were short-lived. The Mukden Incident initiated an army-led colonization process, notably for Manchukuo. From 1931, the increase of overseas Japanese in China was due to those who moved to Manchukuo. The "Migration Plan of 1 million Households and 5 million People" was elaborated by the Kwantung Army and decided by the Cabinet led by Japanese Prime Minister Koki Hirota in 1936. A detailed plan for the first phase aimed to increase 1 million Japanese in Manchukuo for the five years from 1937 to 1941 [36]. Just before the first and last population census of Manchukuo was conducted in October 1940, a fertility survey was carried out in July by the Department of Civil Affairs of Manchukuo, assisted by Dr. Mutsuo Nishio, a research officer of the Institute of Population Problem of Japan [37], by which results, the population projection was conducted [38]. At the start of 1937, the number of Japanese residing in Manchukuo was 385,865, which rose to 1,016,805 at the end of 1941, an increase of 630,940 in the 5 years of the Plan's first phase [39, 40]. It did not reach what was planned. Nevertheless, until 1942, when the last statistics were available, the number of Japanese living in Manchukuo and Kwangtung province increased rapidly to 1,319,599.

During the period, the movement of non-Japanese within the Japanese territories was active, notably the Koreans. The Koreans are very mobile people even now [2], but also back in the pre-war period. 1,241,315 Koreans were counted in the 1940 census of mainland Japan, far more than 22,499 Taiwanese, 986 Sakhalin, or 249 South Seas Islanders. In addition, in the same year, there were 1,450,384 Koreans in Manchukuo, 16,056 in Sakhalin, 5710 in Kwantung Province, 3463 in South Seas Islands and 392 in Taiwan, according to the population censuses and registers of each territory. Within the Imperial Japanese territories, people could move rather easily. However, the move was not always voluntary, and regarding the forcefulness of the migration, the dispute continues between Korea and Japan up to the present.

Although the number of cross-border migrants increased rapidly, it was far less than the number of migrants within the country. The 1928 response to the Commission for the Investigation of Problems of Population and Food clearly stated that internal migration should precede international migration, as it was far more effective [1]. Indeed, towards the end of WWII, there was massive internal migration. The National Service Draft Ordinance was promulgated to relocate people for employment in military-related industries. Then, the evacuation against bombardment was organized especially for the school children, but adult city dwellers did so voluntarily. The direction of evacuation was from urban to rural [41]. For example, the population of Tokyo prefecture decreased by 4 million from 1944 to 1945.

Towards the end of WWII, the population statistics ceased to be published. The last Imperial Japanese Statistics Yearbook was published in February 1941, which contained the population data up to 1939 only [42]. The censuses undertaken throughout the territory in October 1940 were never properly published as one report. The advanced announcement in the official gazette was made on 18th April 1941 on the total number of the imperial territory, summing up to 105,226,101 persons [43]. However, the detailed figures were published after the war for the results of mainland Japan only [44, 45]. The 1940 census reports of overseas territories were published only for Kwantung Province [46], and the reports of Taiwan, Korea, Sakhalin, and South Seas Islands, as well as those conducted in Manchukuo, were not properly published. The existence of statistics is a sign of the sanity of society. The peopling in the Greater East Asia Co-Prosperity Sphere, without proper statistics, was unsustainable.

3.6 Repatriation

With the war over, millions of overseas Japanese had to rush back to the small Japanese mainland. In addition, the Japanese government was responsible for returning non-Japanese, notably Koreans and Taiwanese, back to their home countries. 7.6 million people, composed of 3.2 million civilian overseas Japanese, 3.1 million military Japanese and 1.3 million non-Japanese residing in Japan, moved in and out of Japan within a short period after the end of the war [47, 48]. Upon the order of GHQ (General Headquarters, the Supreme Commander for the Allied Powers), the Ministry of Health and Welfare became responsible for the task, absorbing relevant sections of former military ministries. In addition to the central level, dozens of regional bureaux were created to manage the massive in-and-out migration through the ports authorized by the GHQ.

The military personnel return was stipulated in the ninth paragraph of the Potsdam Declaration that they "shall be permitted to return to their homes with the opportunity to lead peaceful and productive lives". The repatriation process was swift and prioritized over the repatriation of civilian Japanese. As for the civilian return, it was not bound by international agreement, and the Japanese government originally considered those overseas Japanese should stay where they were for the

3.6 Repatriation

sake of their lives and property. However, the civilian return was preferred by GHQ to remove the possible Japanese influences in those places [49]. The repatriation process was rather smooth in the US, UK, Australia, and Chinese Kuomintang-ruled areas and finished by the end of 1947.

However, in the north, the repatriation was stopped due to the advancement of the Soviets and the formation of new authorities, namely the People's Republic of China and the Democratic People's Republic of Korea. The repatriation from the northern part of Korea was stopped in July 1948, former Manchukuo in August 1948, Kwantung Province in October 1949, and Sakhalin and Soviet mainland in April 1950, at which time the Telegram Agency of Soviet Union announced that the return process was completed. It shocked Japan as some 369,000 persons were considered to remain there. The Ad Hoc Commission on Prisoners of War was set up within the United Nations in 1950, and collective repatriation resumed in 1953. With the restoration and normalization of the diplomatic relationship with the USSR in 1956 and the People's Republic of China in 1972, the repatriation process continued, and by the end of 1995, 42,269 persons returned to Japan alive.

Collective repatriation ended in 1959, but individual repatriation continued. As of 1996, 709 persons were listed as not yet returned. The governmental repatriation assistance shifted to the collection of bones of the deceased overseas. Out of 2.4 million military and civilian Japanese who died overseas, 1.2 million bones returned [48].

As for the repatriation of non-Japanese military personnel and conscript labourers, the repatriation started right after the end of the war, and by the end of 1945, almost all returned home. As for the civilian non-Japanese who were residing in Japan, the GHQ decision was made belatedly in March 1946, and they returned home by the same ship used for the Japanese repatriation. The mandate finished by the end of 1950 and the number of repatriations rose to 1.3 million. As of 1st May 1950, among the total of 1,194,362 returnees, 945,771 were Koreans, 180,430 were Ryukyuan people, 65,912 were Chinese, including Taiwanese, and 2249 were of other nationalities, including 2033 Germans and Italians, and 1036 Indonesians. The return of Koreans was the most important and also had difficulties. The actual number of people who showed up at the ports was much smaller than expected. Incidents such as the cholera outbreak in Busan, the railway disruption caused by severe flooding in the south of Korea, and the general strike of railway workers in Korea discouraged the departure from Japan. Also, the political insurgence and atrocity that happened in Cheju island caused a new unauthorized migration to Japan, and among those, 24,622 ended up being deported back again to Korea.

The repatriation was considered the largest mass migration in the history of Japan. Indeed, among the 6.3 million Japanese who returned home, 5 million did so from August 1945 to the end of 1946. The increase of 5 million people within one year has never happened even up to the present. Anticipating the chaotic situation, preparation was needed based on numbers. In August 1946, the Headquarters for Economic Stabilization released the population projection up to 1950. Based on the population survey conducted in April 1946, the population in 1950 was projected to be 79,461,000 [50]. In reality, the census of 1950 counted 83,199,637, around 4 million more than the projection made only 4 years back. The difference was due to an underestimation of births by 1.6 million and an overestimation of deaths by 2.8 million, but not much for international migration. While post-war migration was unprecedented, the concurrent changes in fertility and mortality proved even more unpredictable.

3.7 The Creation of New Migration Statistics and the Surge of Post-War Internal Migration

The households started to be registered by the municipal government and neighbourhood mutual-aid association in around 1940, separately from the family register [51]. To manage the rationing of food and tasks as well as delivery of community services, an accurate register of *de facto* residents was needed—something the *de jure* family register could not provide. While keeping the family register for the sake of authentication of family relationship, resident registration based on the location of residence, was institutionalized by law in 1951. The municipal government maintained it, but the migration information between the municipalities was collected by the national government, which saw the start of new and reliable internal migration statistics. The Report on Internal Migration, as a national statistic, counts the monthly and annual numbers of internal migrants derived from the Basic Resident Registration from 1954 to the present. It is considered to be unique among the internal migration statistics available in the world [52].

The number of internal migrants started to increase in 1960, peaked in 1973, and then gradually decreased until 2013, when the number remained stagnant (Fig. 3.6). The demographic structure explains these trends. The youth, especially in their 20s, tend to move a lot, and their number determines the number of internal migrants. However, the trends do not match exactly. The youth increase in the 1990s did not affect the number of internal migrants much compared to the 1960s. This decreased mobility is explained by lower fertility and decreased "potential lifetime out-migrants" [53]. With the traditional family system, the eldest child was supposed to succeed the family, so they tended to stay. If there were only the eldest son and daughter, the typical situation with a total fertility rate of 2, there would be no need to migrate out. The fertility level determines the relationship between the number of youth and internal migrants.

Apart from these purely demographic factors, mobility was affected by the increasing level of educational achievement, better transport infrastructure, gap in labour demand and supply and the responsive policies. From 1955 to 1965, the more rapid increase of internal migrants compared to the number of youths can be explained by assisted labour placement. The drop in the number of internal migrants in 1973 was partly due to the "Oil Shock", but also the deconcentrating policy with the slogan of "Remodelling the Japanese Archipelago" [54] could have given the impact.

3.8 Labour Placement: From Order, Assistance to Choice 59

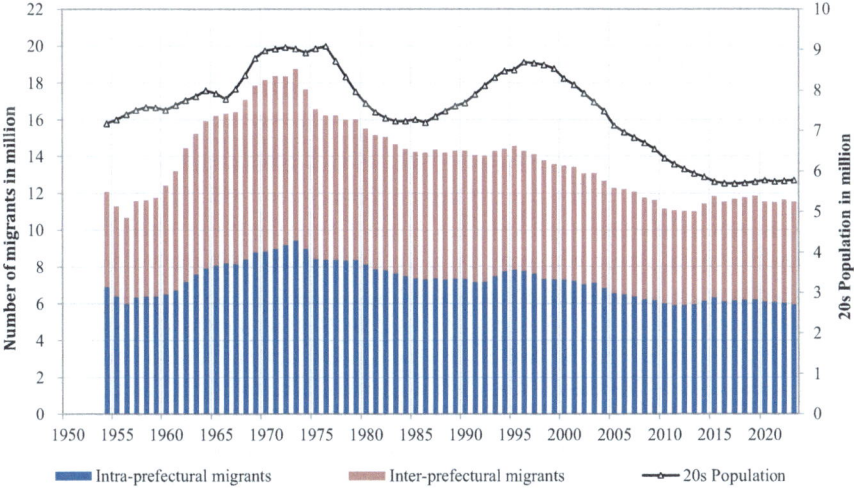

Fig. 3.6 Annual intra and inter-prefecture migration number and rate. Note: Japanese until 2013, Japanese and foreign nationals from 2014. Source: Report on Internal Migration (Statistics Bureau)

The moderate increase and stagnation of internal migrants since 2014 were due to the inclusion of foreign nationals in the statistics as they started to be included in the resident registration in 2012. The foreign nationals are young, mobile, and increasing, but so far, the number has not been at the level to increase the total number of internal migrants drastically.

3.8 Labour Placement: From Order, Assistance to Choice

The long-distance employment placement policy, which requires migration, was implemented firstly for the textile female worker back in 1927. That was the time when Japan, a member of ILO since its creation in 1919, ratified the Unemployment Convention of 1919, set up the public employment agency, and stipulated the Employment Placement Act in 1921. It carried out its mission in collaboration with existing private and cooperative networks [55].

Following the start of the China-Japan War, the National Mobilization Law was stipulated in 1938, and labour mobility became strictly under the control of the national government to proceed with the war. Related laws and orders were implemented which enforced the registration of skilled labour and strengthened the jurisdiction of the governmental employment agency [47, 56].

Although Japan lost 3 million people killed in the war, the repatriation of Japanese from former territories, followed by the post-war baby boom, instantly replaced the loss. The baby boom and a rapid decline of mortality caused the population to surge by 11 million in just 5 years from 1945 to 1950. Combined with post-war chaos, the immediate concerns were overpopulation, poverty and unemployment. The Advisory Council of Population Problems (ACPP), set up in 1949, quickly proposed policies to ease unemployment, such as public unemployment insurance and a minimum wage system. Although this Prime Ministerial Advisory Council was restructured to be placed under the Ministry of Health and Welfare in 1953, the renewed ACPP considered the labour issue as one of the primary issues to solve the population problems. It adopted the Resolution on Population Carrying Capacity in 1955 [57]. While the registered unemployment rate was not so high, around 2% of the labour force, the "potential unemployment", those who were employed but underpaid in the middle or small sized enterprise, in agriculture and day labourer, became the focus of attention. In 1958, the Resolution on the Measures against Potential Unemployment was adopted by ACPP [58], and economic and employment policy and minimum wage system were proposed again. Minimum Wage Act was enacted the following year.

Among those policies, "rational mobility of labour force" was highlighted. The post-war implementation of prolonged compulsory education up to junior high school for those baby boomers born from 1947 to 1949 created a surge of the young labour force in search of jobs upon junior high school graduation at age 15. While they faced potential unemployment in rural agricultural areas, substantial labour shortages were emerging in metropolitan areas. This regional disparity led to the amendment of the Employment Security Act in 1960 to enable long-distance employment placement. In 1962, the Ministry of Labour planned to relocate 6000 job-seekers in rural areas to Tokyo, Osaka and Nagoya metropolitan areas, and the number of those actually moved rose to 12,760 persons, more than double than planned [59]. Special trains were organized to carry those young workers from rural to urban, notably from the Tohoku area to Tokyo. The total number of junior high school graduates who moved across the prefectural border rose from 100,000 in 1953 to almost 250,000 in 1963 [60].

The job entry of many junior school graduates who searched for jobs was, however, a short-lived event. Those who continued to high school and university increased thereafter (Fig. 3.7). Accordingly, the number of employed shifted from junior high school graduates to high school graduates, then university graduates (Fig. 3.8). The policy efforts for labour placement for new graduates continued, but the long-distance employment assistance became less important.

3.9 From a Sending to a Receiving Nation: The Transformation of International Mobility

The post-war international mobility in Japan "increased rapidly," according to contemporary official reports [61], but remained very low compared to the pre-war period or other countries. Overseas Japanese plummeted from prewar 3.3 million to 0.2 million, less than one-tenth (Fig. 3.1). As for the foreign nationals, many Koreans, Chinese and Taiwanese returned to their country voluntarily or by deportation. For example, among the 2,049,561 Koreans in Japan at the end of the WWII, 1,023,459 persons returned voluntarily (from 1945 to 1950, [62]), and 3633 were

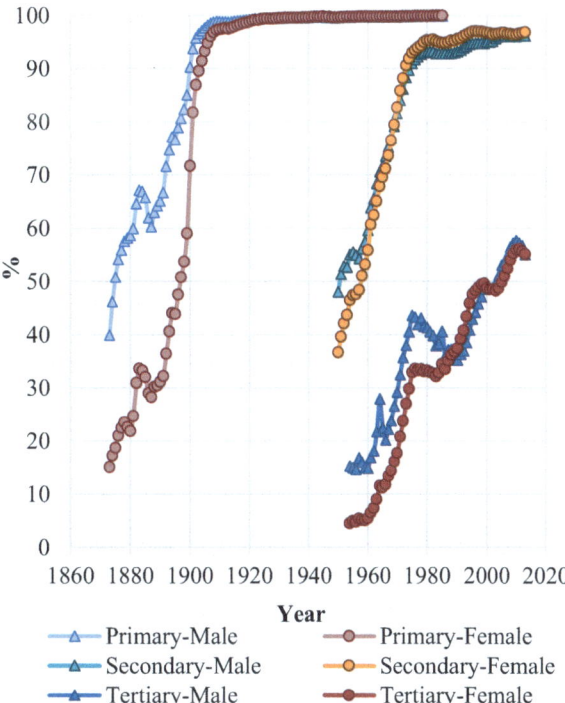

Fig. 3.7 Educational coverage rate. Source: School Basic Survey (Ministry of Education, Cultures, Sports, Science and Technology)

deported by newly created Immigration Bureau (from 1950 to 1952, [61]). The number of non-Japanese fell significantly from a pre-war high level to 0.6 million by 1948. The number increased with the additional newcomers but only to 0.8 million at the end of the 1980s. Most of them were Koreans who remained in Japan from before WWII (Fig. 3.9).

The WWII devastation brought the Japanese economy to the nadir, and the sudden increase in population, by repatriation and baby boom, revived the fear of overpopulation. Among the three options to ease the population pressure, namely economic development, emigration and birth control, the policy on emigration was not feasible at the time as Japan was under the GHQ administration, which feared Japanese revival in Asia [49] and Japan was without its own diplomatic authority. After the independence in 1952, the out-migration policy resumed. The cargo liner Santos Maru left Kobe port as early as December 1952 to dispatch the first post-war migrant to Latin America.

In fact, the promotors of post-war Japanese out-migration were the Americans. Prime Minister Shigeru Yoshida, who saw the economic boom in Italy due to the emigrants' remittance from the US, thought it would be a good idea to invest in emigration and agreed to receive the loan from three American banks, National City, Chase Manhattan and Bank of America, mediated by Mr. Rockefeller in 1954 [63]. Based on the loan, the Japan Emigration Promotion Co. Ltd. was created and promoted the out-migration of Japanese. At the same time, the Bureau of Immigration and Immigration Council was inaugurated in 1955.

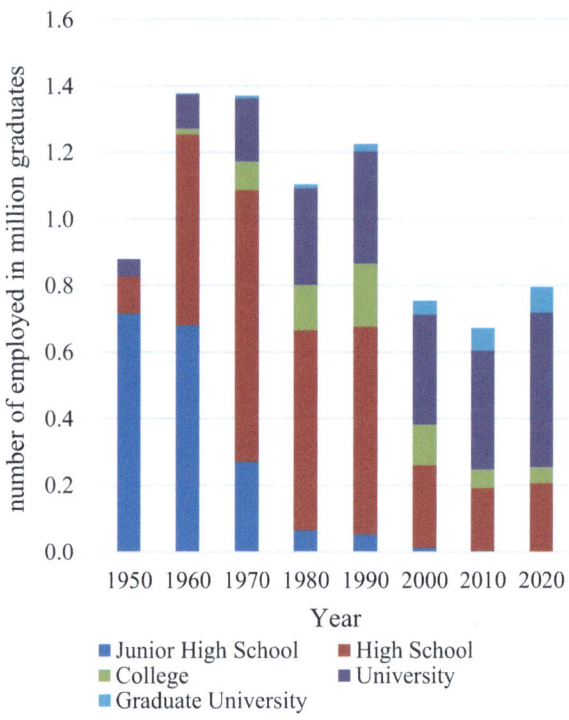

Fig. 3.8 Number of employed upon graduation. Source: School Basic Survey (Ministry of Education, Cultures, Sports, Science and Technology)

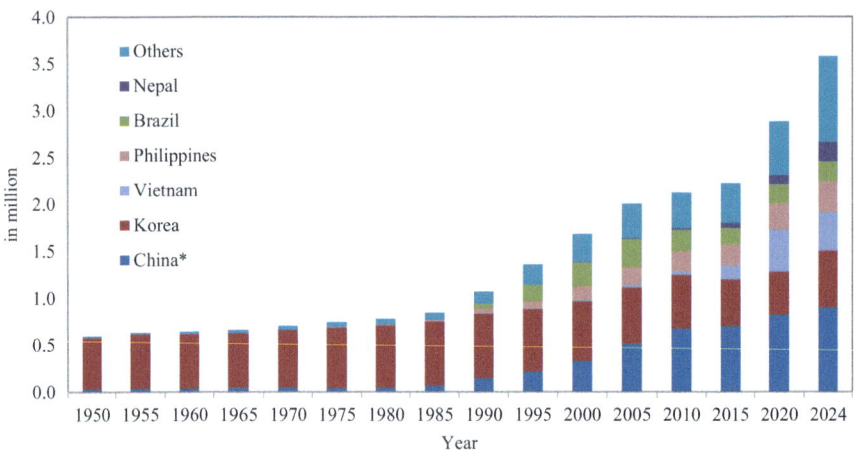

Fig. 3.9 Number of registered foreign nationals by nationality (1950–2024). Note: China includes Taiwan. 2024 is at the end of June. Source: Immigration Services Agency

The number of emigrants increased to 16,620 in 1957 but kept on decreasing since then. Emigration never again reached pre-war levels. Instead, the overpopulation was absorbed by the internal migration to supply the demand for labour during the high economic growth of the 1960s [64]. Yen, the Japanese currency, literally meaning "circle", was fixed to 360 yen to one US dollar, as there are 360 degrees in a circle, making overseas emigration too expensive for the Japanese. Also, it became difficult to clearly define "emigrant". Air travel became more common, which allowed people to move back and forth more frequently. Since 1964, international travel has been allowed to all Japanese, and people use the passport multiple times until the expiration date. At the time of departure, one did not have to decide whether the trip was a lifetime choice or a temporary stay. The number of international travellers soared so did the number of overseas Japanese (Fig. 3.1).

It was in 1982 when the GHQ legacy order was reformed into proper legislation in the form of the Immigration Control and Refugee Recognition Act. New types of visas for technical intern training and descendants of overseas Japanese were created, which boosted the speed of the increase of foreign nationals. The global financial crisis in 2008 affected Japan, like other high-income countries. The number of new entries decreased, and many foreign nationals, especially Nikkei Brazilians, were escorted to return back, receiving some governmental subsidy to purchase the return ticket as well as some stipend. When the other high-income countries regained the number of immigrants in the following years, Japan did not, as the Great East Japan Earthquake hit it on 11th March 2011. Many feared the radioactivity and moved out of Japan. However, this downturn was reversed quickly in 2013, and the increase of foreign nationals continued until the COVID-19 pandemic.

3.10 Migration Policies in an Era of Population Decline

Population structure and trend are both causes and consequences of migration. As the population ages, the proportion of young and mobile people decreases; hence, the internal mobility rate declines. However, the exodus of youth from non-metropolitan to metropolitan areas continues, notably at times of major life events such as entering university and getting a job. Over the course of population ageing and decline, depopulation has been a serious matter in rural, non-metropolitan areas. The policies to mitigate the depopulation were initiated with the Emergency Measures Law for Depopulated Areas, came into force in 1970. Since then, measures have been taken, and some policies in the 1970s successfully slowed the migration to metropolitan areas, as explained before. Turning into the twenty-first century, when the population decline at the national level started and became apparent, the depopulation, Tokyo monopolization and rural municipality extinction became a major policy focus. The Act on Overcoming Population Decline and Vitalizing Local Economy in Japan was stipulated in 2014, and a dedicated bureau in the national government implemented policies to reverse the people's movement back to the local area. However, it was difficult to change the free movement of people. Since 2014, the annual net migration of the Tokyo area (Tokyo, Chiba, Saitama and Kanagawa prefectures) has never gone negative, meaning that there are always more people coming in than leaving, even during the COVID-19 period.

It is difficult to label the 2014 Act as a failure, as nobody knows what would have happened without it. It could have been much worse without the policies implemented by the Act. In fact, not only the Tokyo area but some other urban prefectures or large cities are gaining population. The continuous positive net migration has been found in Osaka and Fukuoka at the prefectural level since 2015 and in Sapporo, Sendai, Nagoya, Osaka, and Fukuoka at the city level since 2014. What is certain is that urbanization is an inevitable trend, and the hierarchy of central cities is becoming more apparent.

On the other hand, along with the total population decline, the rural, remote "underpopulated areas", to which development measures were taken since 1970, are increasing its number, but they are somewhat more resistant than expected. According to the survey on communities in underpopulated areas conducted since 1999, the extinction of underpopulated areas occurred only around 70% of those estimated 10 years before [65]. The policies including infrastructure expansion, especially road and mobile phone networks, as well as human support, seem to be helping community disappearance.

Since 2012, Japanese international migration policy entered a new era. That was the year when the foreign nationals were integrated into the resident registration. The annual governmental Basic Policies expanded the scope of policies year by year for introducing and integrating foreign residents [66]. The target shifted from international students and travellers to foreign high-skill professionals, then to middle-skill workers. In 2018, the Comprehensive Measures for Acceptance and Coexistence of Foreign Nationals, along with a five-year roadmap for better social integration, started to be implemented. COVID-19 suppressed this exploding trend but soon resumed in 2022. In 2024, the proportion of foreign nationals among the total population is still 2.7%, but it is projected to rise to 10.7% in 2070 [67]. This figure is still modest compared to the present level of Western countries, but Japan will certainly be a country with a more diverse culture.

Internal and international migration are major determinants of population. As in the case of fertility, some policies are effective, and some are not, in controlling the movement of people. The political decision should be made based on sound data. Policies are affected by value but also create new value.

References

1. CIPPF: Commission for the Investigation of Problems of Population and Food. 1931. *Jinkō shokuryō mondai chōsakai yōran (Directory of the Commission for the investigation of problems of population and food)*. Tachi Archive no. 10121607.
2. Hayashi, Reiko. 2014. International comparison of migration - A construction of model-mobility using Japanese indicators. *Journal of Population Problems (Jinkō Mondai Kenkyū)* 70 (1): 1–20. https://www.ipss.go.jp/syoushika/bunken/data/pdf/19917901.pdf.

References

3. Hayashi, Reiko. 2021. Overseas Japanese statistics in pre-war period. *Journal of Population Problems (Jinkō Mondai Kenkyū)* 77 (3): 259–265. https://doi.org/10.50870/00000287.
4. Japan Statistical Association. 1988. *Historical statistics of Japan*. Supervised by Statistics Bureau.
5. Tōkeiin (Statistics Bureau). 1885. *Dai Yonkai Nihon Teikoku Tōkei Nenkan (The 4th statistical yearbook of imperial Japan)*.
6. Statistics Bureau of the Cabinet. 1938. Shōwa 5 nen Kokusei Chōsa Saishū Houkokusho (Final Report of the 1930 Population Census). https://dl.ndl.go.jp/pid/1281995.
7. Kaempfer, Engelbert, translated by John Gaspar Scheuchzer. 1906. *The history of Japan: Together with a description of the kingdom of Siam 1690-92*. Glasgow: James MacLehose and Sons. https://archive.org/details/historyofjapanto02kaem/page/n7/mode/2up
8. Shizuki, Tadao. 1801. *Sakoku ron yaku rei (Translation of locked country theory)*. Katsu Kaishu Kankei bunsho 88. http://dl.ndl.go.jp/info:ndljp/pid/11222542
9. Sugimoto, Tsutomu. 2015. *Sakoku Ron - Inei honkoku kōchu (Locked country theory - Imprint, transcription, annotation)*. Yasaka Shobo.
10. Hakone Sekisho. n.d. Zenkoku no sekisho (Checkpoints in the nation). https://www.hakone-sekisyo.jp/db/data_inc/inc_frame/fr_data_01_01_01.html
11. Kanemitsu, Hideo. 2006. The seclusion policy (Sakoku) of Tokugawa government and it's significance of contemporary problems. *The Journal of Island Studies* 2006 (6). https://doi.org/10.5995/jis.2006.1
12. Kito, Hiroshi. 2002. *Bunmeitoshiteno Edo system (The Edo system as a civilization)*. Kodansha.
13. Takebe, Kenichi. 2015. *Dōro no nihonshi - Kodai ekiro kara kōsoku dōro e (The history of roads in Japan: From ancient station paths to automobile highways)*. Chuokoron-Shinsha.
14. Hayami, Akira. 1992. *Population, economy and society in early modern Japan: A study of the Nobi Region*. Sobunsha.
15. Kumada Tadao. 2010. Sokoni Nihonjin ga ita! - Umi wo watatta gosenzosamatachi (There were Japanese people there! - Our ancestors who crossed the sea). Shinchobunko Ku-36-1.
16. Watsuji, Tetsuro. 1950. *Sakoku - Nihon no higeki (Locked country - The tragedy of Japan)*. Chikuma Shobo. https://dl.ndl.go.jp/pid/1663925/1/5.
17. Cabinet Official Gazette Bureau. 1888. Meiji 4 Hourei Zensho (Compendium of Laws). https://dl.ndl.go.jp/pid/787951
18. Saito, Osamu. 1973. Population migration before 1920: A preliminary study based on the Kiryu statistics of four prefectures. *Mita Gakkai Zasshi (Mita Journal of Economics)* 66 (7): 500(56)–508(64).
19. Statistics Bureau of the Cabinet. 1913. Ishin igo teikoku tōkei zairyō shūsan. Dai 4 hen (Jinkōdōtai ni kansuru tōkei Zairyō) (Collection of imperial statistical materials after the Meiji Restoration, Part 4 (Statistical materials related to vital statistics)). https://dl.ndl.go.jp/pid/946279/1/107
20. Department of Geography, Ministry of Interior. 1885. Meiji 17 nen 8 gatsu 9 gatsu bōfū kiji (The report of windstorm in August September of Meiji 17). https://dl.ndl.go.jp/info:ndljp/pid/831933
21. Yasuda, Taijiro. 1941. *Hokkaido imin seisaku shi (History of Hokkaido immigration policy)*. Seikatsusha. https://dl.ndl.go.jp/pid/1269987.
22. Uehara, Tetsusaburo. 1914. *Hokkaidō Tondenhei Seido (Hokkaido Tondenhei System)*. Hokkaido Government. https://dl.ndl.go.jp/pid/942119.
23. Mayet, Paul. 1888. *Landwirthschaftliche versicherung*. Tokyo: Kokubunsha. https://dl.ndl.go.jp/pid/1675622/1/7. Japanese translation https://dl.ndl.go.jp/pid/802484.
24. Lu, Sidney Xu. 2019. Eastward Ho! Japanese settler Colonialism in Hokkaido and the making of Japanese migration to the American West, 1869–1888. *The Journal of Asian Studies* 78 (3): 521–547. https://www.jstor.org/stable/10.2307/26758246.
25. Yoshida, Hideo. 1944. *Nihon jinko ron no shiteki kenkyu (Historical research on Japanese population theory)*. Kawade Shobo. https://dl.ndl.go.jp/pid/1061472
26. Kikuchi, Yoshiki. 1981. The study of internal migration - Population changes in Hokkaido: 1869-1925. *Studies in Sociology, Psychology and Education* 21:21–34. https://koara.lib.keio.ac.jp/xoonips/modules/xoonips/detail.php?koara_id=AN0006957X-00000021-0021.

27. Watanabe, Susumu. 1994. The Lewisian turning point and international migration: The case of Japan. *Asian and Pacific Migration Journal* 3 (1): 119–147. https://doi.org/10.1177/011719689400300107.
28. Ministry of Foreign Affairs of Japan. n.d. Gaikō Shiryō Q&A Sonota Pasupōto Kankei (Diplomatic materials Q&A other, related to passport). https://www.mofa.go.jp/mofaj/annai/honsho/shiryo/qa/sonota_01.html
29. Irie Toraji. 1942. *Hōjin kaigai hatten shi (History of overseas Japanese development)*. Ida Shoten. https://dl.ndl.go.jp/pid/1461457
30. IPP: Institute of Population Problems. 1942. *Hōjin kaigai hatten shi ryakusetsu (Brief description of the history of overseas Japanese development)*. Daitōa Kensetsu Minzoku Jinkō Shiryō (Great East Asia construction race and population material) No. 38, 42, 43, 44. https://www.ipss.go.jp/syoushika/bunken/data/pdf/J08660.pdf. https://www.ipss.go.jp/syoushika/bunken/data/pdf/J08671.pdf. https://www.ipss.go.jp/syoushika/bunken/data/pdf/J08670.pdf. https://www.ipss.go.jp/syoushika/bunken/data/pdf/J08664.pdf
31. JICA: Japan International Cooperation Agency. 1994. *Kaigai ijū tōkei (Showa 27 nendo - Heisei 5 nendo) (International Migration Statistics (FY1952–1993))*. Gyomu Shiryo (Business Material) No. 891. https://www.jica.go.jp/Resource/jomm/outline/library/ku57pq00000lx70u-att/statistics.pdf
32. Tōkeiin (Statistics Bureau). 1882. *Tōkei Nenkan (Statistical yearbook)*.
33. Kurahashi, Masanao. 1989. *Kita no karayuki san (Japanese female emigrant to the north)*. Kyoei Shobo. https://dl.ndl.go.jp/pid/13150332
34. Boo, Hak Joo. 2006. Comparison and chronological rearrangement of historical plans for Japanese Colony (Soryo Wakan' in Busan in early modern ages). *Transactions of AIJ* 609:147–154.
35. Davidson, James W. 1903. *The Island of Formosa - Past and present. History, people, resources, and commercial prospects*. Macmillan, Kelly & Walsh.
36. Ministry of Colonial Affairs. 1937. Manshu imin daiikki keikaku jisshi yōryō (Manchuria emigration first phase plan implementation guidelines). https://dl.ndl.go.jp/pid/1441968
37. Manchukuo Department of Civil Affairs, Welfare Division. 1941. *Manshu koku jinkō mondai kenkyū dai isshū (Manchukuo population problem research)*. Vol. 1. Tachi Archive.
38. Manchukuo Bureau of Planning. 1941. *Manshu koku Shōrai Jinkō no Suikei - Jinkō Haichi Keikaku Kenkyū (Sono Ichi) (Population projection of Manchukuo - Research on population allocation plan (part 1)*. Tachi Archive TACHIB99OM4108
39. Kwantung Bureau. 1938. Shōwa 12 nen Kantō kyoku kannai genjū jinkō tōkei (Showa 12th year Kwantung Bureau territory current population statistics). https://dl.ndl.go.jp/info:ndljp/pid/1451690
40. Kwantung Bureau. 1944. Shōwa 18 nen Kantō kyoku kannai genjū jinkō tōkei (Showa 18th year Kwantung Bureau territory current population statistics). https://dl.ndl.go.jp/info:ndljp/pid/1439947
41. Tani, Kenji. 2012. A cohort analysis of internal migration in Japan during the 1940s. *Geographical Review of Japan Series A* 85 (4): 324–341. https://doi.org/10.4157/grj.85.324.
42. Statistics Bureau of the Cabinet. 1941. The 59th statistical yearbook of the great empire of Japan. https://dl.ndl.go.jp/info:ndljp/pid/1275672
43. Printing Bureau of Finance Ministry. 1941. Teikoku Zenhanto no Jinkō (Population of entire territory of the empire). Japan Official Gazette, 18th April 1941. https://dl.ndl.go.jp/pid/2960780
44. Statistics Bureau. 1949. Results report summary of 1940 population census, 1944 population survey, 1945 population survey and 1946 population survey. https://dl.ndl.go.jp/pid/9524279/1/4
45. Statistics Bureau. 1961. 1940 population census of Japan
46. Kwantung Bureau. 1941. Shōwa 15 nen Kantō Shū Kokusei Chōsa Setai Oyobi Jinkō (1940 Kwantung Province population census households and population). https://dl.ndl.go.jp/pid/1276946

47. Ministry of Health and Welfare Editorial Committee for 50 years History of Ministry of Health and Welfare. 1988. *Kōseishō 50 nen shi (50 years history of Ministry of Health and Welfare)*. Foundation-Institute for Research of Health and Welfare Problems.
48. Ministry of Health and Welfare. 1997. *Engo 50 Nen Shi (50 years history of war victims' relief)*. Gyōsei.
49. GHQ/SCAP, commentary by Toshio Kuroda, translated by Toshio Kuroda and Michiko Obayashi. 1996. History of the non-military activities of the occupation of Japan, 1945–1951, Volume 4, Population. Nihon Tosho Center.
50. Headquarters for Economic Stabilization. 1946. *Shōrai jinkō no suikei ni kansuru hōkoku (Report regarding the future population projection)*. Statistics Research Group Population Sub-Group. https://dl.ndl.go.jp/pid/1267690
51. Kaji, Ryosaku. 1951. *(On the draft resident registration law)*. 10th diet, house of representatives, committee on judicial affairs, No. 15. https://kokkai.ndl.go.jp/txt/101005206X01519510327/16
52. Tachi, Minoru. 1960. *Formal demography*. Tokyo: Kokon Shoin Co.
53. Itoh, Tatsuya. 1984. Recent trends of internal migration in Japan and "population life time out-migrants". *Journal of Population Problems (Jinkō Mondai Kenkyū)* 172:24–38. https://www.ipss.go.jp/syoushika/bunken/data/pdf/14168502.pdf.
54. Tanaka, Kakuei. 1973. *Nihon rettō kaizō Ron (Remodeling the Japanese archipelago)*. Nikkan Kogyo Shimbun.
55. Kanbayashi, Ryo. 2009. Naze Shokugyō Shōkai wa Kuni ga Okonaunoka (Why employment placement is conducted by the state?). *The Japanese Journal of Labour Studies (Nihon Rodo Kenkyu Zasshi)* 585:66–69.
56. Ministry of Labour. 1961. *Rōdō gyōsei shi*. Vol. 1. https://dl.ndl.go.jp/pid/9524869
57. ACPP: Advisory Council of Population Problems. 1955. Jinkō shūyōryoku ni kansuru ketsugi (Resolution on population carrying capacity). https://www.ipss.go.jp/history/shingikai/data/J000008913.pdf https://www.ipss.go.jp/syoushika/bunken/data/pdf/14206803.pdf
58. ACPP: Advisory Council of Population Problems. 1958. Senzai shitsugyō taisaku ni kansuru ketsugi (Resolution on the measures against potential unemployment). https://www.ipss.go.jp/history/shingikai/data/103889.pdf https://www.ipss.go.jp/syoushika/bunken/data/pdf/14207804.pdf
59. Ministry of Labour. 1973. *Rōdōshō 25 Nen Shi (25 years history of Ministry of Labour)*. Rōdō Gyōsei Chōsa Kenkyūkai. https://dl.ndl.go.jp/pid/12008818.
60. Kariya, Takehiko, Shinji Sugayama, and Hiroshi Ishida. 2000. *Gakkō shokuan to rōdō shijō - Sengo shinki gakusotsu shijō no seidoka katei (Schools, employment offices and the labor market - The process of institutionalization of the postwar new graduate market)*. The University of Tokyo Press.
61. Ministry of Justice, Immigration Bureau. 1959. *Shutsunyukoku Kanri to Sono Jittai (Immigration control and its situation)*.
62. Ministry of Justice, Immigration Bureau. 1953. *Sūji kara mita zainichi Chosenjin (Korean residents in Japan by numbers)*. Nyukan Shitsumu Chōsa Shiryō No. 8.
63. Ministry of Foreign Affairs, Consular and Migration Affairs Division. 1971. Waga kokumin no kaigai hatten - Ijū hyakunen no ayumi (Honpen) (The overseas development of Japanese - A hundred years of migration (main volume)). https://dl.ndl.go.jp/pid/11969926
64. Kajita, Takamichi. 1992. *Gaikokujin rōdōsharon: Genjō kara riron e*. Kobundo.
65. Kasotakisakushitsu (Division of Depopulation Countermeasures, Group for Regional Vitalization, Ministry of Internal Affairs and Communications). 2020. Kasochiikitouni okeru shūrakuno jōkyōni kansuru genkyōhaaku chōsa houkokusho (Report of the survey on current status of settlements in depopulated areas). https://www.soumu.go.jp/menu_news/s-news/01gyosei10_02000066.html
66. Cabinet Office. 2013–2024. Basic policy on economic and fiscal management and reform. https://www5.cao.go.jp/keizai-shimon/kaigi/cabinet/honebuto/honebuto-index.html
67. National Institute of Population and Social Security Research. 2023. *Population projections for Japan: 2021–2070 (With long-range population projections: 2071–2120)*. Population Research Series, No. 347. https://www.ipss.go.jp/history/foundation/PDF/J000003063.pdf

The manufacturer's authorised representative in the EU is Springer Nature Customer Service Centre GmbH, Europaplatz 3, 69115 Heidelberg, Germany. If you have any concerns regarding our products, please contact ProductSafety@springernature.com

Printed and bound by CPI Group (UK) Ltd, Croydon, CR0 4YY

23/03/2026

02076360-0002